NATIONAL ACADEMIES Sciences Engineering Medicine

NATIONAL ACADEMIES PRESS
Washington, DC

Nuclear Terrorism

Assessment of U.S. Strategies To Prevent, Counter, and Respond to Weapons of Mass Destruction

Committee on Assessing and Improving Strategies for Preventing, Countering, and Responding to Weapons of Mass Destruction Terrorism: Nuclear Threats

Nuclear and Radiation Studies Board

Division on Earth and Life Studies

Consensus Study Report

NATIONAL ACADEMIES PRESS 500 Fifth Street, NW Washington, DC 20001

This activity was supported by a contract between the National Academy of Sciences and U.S. Department of Defense. Any opinions, findings, conclusions, or recommendations expressed in this publication do not necessarily reflect the views of any organization or agency that provided support for the project.

International Standard Book Number-13: 978-0-309-70868-5
International Standard Book Number-10: 0-309-70868-0
Digital Object Identifier: https://doi.org/10.17226/27215
Library of Congress Control Number: 2024943245

This publication is available from the National Academies Press, 500 Fifth Street, NW, Keck 360, Washington, DC 20001; (800) 624-6242 or (202) 334-3313; http://www.nap.edu.

Cover credit: U.S. Air Force photo by André Askew

Printed in the United States of America.

Suggested citation: National Academies of Sciences, Engineering, and Medicine. 2024. *Nuclear Terrorism: Assessment of U.S. Strategies to Prevent, Counter, and Respond to Weapons of Mass Destruction*. Washington, DC: The National Academies Press. https:/doi.org/10.17226/27215.

The **National Academy of Sciences** was established in 1863 by an Act of Congress, signed by President Lincoln, as a private, nongovernmental institution to advise the nation on issues related to science and technology. Members are elected by their peers for outstanding contributions to research. Dr. Marcia McNutt is president.

The **National Academy of Engineering** was established in 1964 under the charter of the National Academy of Sciences to bring the practices of engineering to advising the nation. Members are elected by their peers for extraordinary contributions to engineering. Dr. John L. Anderson is president.

The **National Academy of Medicine** (formerly the Institute of Medicine) was established in 1970 under the charter of the National Academy of Sciences to advise the nation on medical and health issues. Members are elected by their peers for distinguished contributions to medicine and health. Dr. Victor J. Dzau is president.

The three Academies work together as the **National Academies of Sciences, Engineering, and Medicine** to provide independent, objective analysis and advice to the nation and conduct other activities to solve complex problems and inform public policy decisions. The National Academies also encourage education and research, recognize outstanding contributions to knowledge, and increase public understanding in matters of science, engineering, and medicine.

Learn more about the National Academies of Sciences, Engineering, and Medicine at **www.nationalacademies.org**.

Consensus Study Reports published by the National Academies of Sciences, Engineering, and Medicine document the evidence-based consensus on the study's statement of task by an authoring committee of experts. Reports typically include findings, conclusions, and recommendations based on information gathered by the committee and the committee's deliberations. Each report has been subjected to a rigorous and independent peer-review process and it represents the position of the National Academies on the statement of task.

Proceedings published by the National Academies of Sciences, Engineering, and Medicine chronicle the presentations and discussions at a workshop, symposium, or other event convened by the National Academies. The statements and opinions contained in proceedings are those of the participants and are not endorsed by other participants, the planning committee, or the National Academies.

Rapid Expert Consultations published by the National Academies of Sciences, Engineering, and Medicine are authored by subject-matter experts on narrowly focused topics that can be supported by a body of evidence. The discussions contained in rapid expert consultations are considered those of the authors and do not contain policy recommendations. Rapid expert consultations are reviewed by the institution before release.

For information about other products and activities of the National Academies, please visit www.nationalacademies.org/about/whatwedo.

COMMITTEE ON ASSESSING AND IMPROVING STRATEGIES FOR PREVENTING, COUNTERING, AND RESPONDING TO WEAPONS OF MASS DESTRUCTION TERRORISM: NUCLEAR THREATS

STEPHEN E. FLYNN (*Chair*), Northeastern University
MADELYN R. CREEDON (*Vice Chair*), The George Washington University
JULIE A. BENTZ, U.S. Army (retired)
MICHAEL DUNNING, Lawrence Livermore National Laboratory (retired)
ROBERT C. DYNES, University of California, San Diego
STEVEN A. FETTER, University of Maryland, College Park
WILLIAM H. GOLDSTEIN, Lawrence Livermore National Laboratory (*until May 5, 2023*)
EDWARD B. HELD, Sandia National Laboratories (retired, *until May 19, 2022)*
ELEANOR MELAMED, National Nuclear Security Administration (retired)
BRENDAN G. MELLEY, National Defense University
SCOTT ROECKER, Nuclear Threat Initiative
JESSICA STERN, Center for Naval Analyses (*as of June 15, 2022*)
KRISTINE L. SVINICKI, University of Michigan
RODNEY K. WILSON, Sandia National Laboratories (retired)

Technical Consultant

NICKOLAS ROTH, Nuclear Threat Initiative

Staff

MICHAEL T. JANICKE, Study Director
CHARLES FERGUSON, Senior Board Director
JENNY HEIMBERG, Senior Program Officer
AYANNA LYNCH, Research Assistant (*until July 11, 2023*)
KAYANNA WYMBS, Research Assistant

NUCLEAR AND RADIATION STUDIES BOARD

BOARD ON LIFE SCIENCES

COMMITTEE ON INTERNATIONAL SECURITY AND ARMS CONTROL

RAYMOND JEANLOZ (*Chair*), University of California
ANDREW ALLEYNE, University of Minnesota
LINTON BROOKS, independent consultant
MARIANA BUDJERYN, Harvard University
MATTHEW BUNN, Harvard University
NANCY CONNELL, John Hopkins Center for Health Security
LINDA ELKINS-TANTON, Arizona State University
STEVEN FETTER, University of Maryland, College Park
DAVID FRANZ, Midwest Research Institute
RICHARD GARWIN, IBM Thomas J. Watson Research Center
ROSE GOTTEMOELLER, Stanford University
DIANE GRIFFIN, John Hopkins Bloomberg School of Public Health
MARGARET HAMBURG, Nuclear Threat Initiative
JOHN HILDEBRAND, University of Arizona
MORIBA JAH, University of Texas at Austin
ALASTAIR JOHNSTON, Harvard University
ROBERT LATIFF, R. Latiff Associates
JAMES LEDUC, University of Texas Medical Branch at Galveston
JEFFREY LEWIS, Middlebury Institute of International Studies at Monterey
HERBERT LIN, Stanford University
DOUGLAS LOVERRO, Loverro Consulting, LLC
RICHARD MESERVE, Carnegie Institution for Science
RICHARD MIES, The Mies Group, Ltd.
PETER PALESE, Icahn School of Medicine at Mount Sinai
LORA SAALMAN, Stockholm International Peace Research Institute
VICTORIA SAMSON, Secure World Foundation
RACHEL SEGALMAN, University of California, Santa Barbara
JIM TIMBIE, Stanford University

Staff

MICAH LOWENTHAL, Director
RITA GUENTHER, Senior Program Officer
HOPE HARE, Administrative Assistant
CANDANCE HUNTINGTON, Research Associate
MARIE KIRKEGAARD, Program Officer (*until 2022*)
BENJAMIN RUSEK, Senior Program Officer
ALEX TEMPLE, Program Officer

BOARD ON CHEMICAL SCIENCES AND TECHNOLOGY

Reviewers

This Consensus Study Report was reviewed in draft form by individuals chosen for their diverse perspectives and technical expertise. The purpose of this independent review is to provide candid and critical comments that will assist the National Academies of Sciences, Engineering, and Medicine in making each published report as sound as possible and to ensure that it meets the institutional standards for quality, objectivity, evidence, and responsiveness to the study charge. The review comments and draft manuscript remain confidential to protect the integrity of the deliberative process.

We thank the following individuals for their review of this report:

DAVID E. ASPNES (NAS), North Carolina State University
DANIEL BLUMENTHAL, U.S. Department of Energy
PETER BOYNTON, Northeastern University
BROOKE BUDDEMEIER, Lawrence Livermore National Laboratory
MICHAEL BURNS, TechSource, Inc.
MICHAEL DINE (NAS), University of California Santa Cruz
CHRISTOPHER FORD, Missouri State University, The Hoover Institution
TODD C. HELMUS, RAND Corporation, Pardee RAND Graduate School
JOHN M. HOLMES, STAX Engineering
MAJA LEHNUS, Private Consultant
DAVID LUCKEY, RAND Corporation
LINDA LOURIE, WestExec Advisors
CHESLAN SIMPSON, Pacific Northwest National Laboratory

Although the reviewers listed above provided many constructive comments and suggestions, they were not asked to endorse the conclusions or recommendations of this report nor did they see the final draft before its release. The review of this report was overseen by **SALLIE ANN KELLER (NAE),** University of Virginia, and **GRANT H. STOKES (NAE),** MIT Lincoln Laboratory. They were responsible for making certain that an independent examination of this report was carried out in accordance with the standards of the National Academies and that all review comments were carefully considered. Responsibility for the final content rests entirely with the authoring committee and the National Academies.

Acknowledgments

Many individuals, organizations, and U.S. government agencies contributed to the successful completion of this report. The committee and National Academies' staff appreciate their kind assistance. Presenters for the numerous briefings are listed in appendix A. Volunteering their time to give valuable insight was crucial to the information gathered and compiled in this report.

The committee would like to give a special thanks to colleagues who helped arrange the meetings for this challenging project spanning many topics, domestic and international, and provided documents and reports necessary to assess the landscape of counterterrorism especially in a rapidly changing world confronted by current events. Those colleagues are:

- *National Nuclear Security Administration*: Jay Tilden, Dallas Boyd, Kasia Mendelson, and Victor Jones
- *Department of Homeland Security*: Gary Rasicot, Michael Kangior, William Bilicic, Frederick Breaux, and Gregory Abide
- *Nuclear Regulatory Commission*: Cynthia Jones
- *Embassy of Canada, Washington, DC*: Jodie McGrath
- *National Security Council*: Elizabeth Sherwood-Randall, Pranay Vaddi, and Brooke Samples

Preface

"The most important failure was one of imagination. We do not believe leaders understood the gravity of the threat. The terrorist danger from Bin Ladin (sic) and al Qaeda was not a major topic for policy debate among the public, the media, or in the Congress.[...] Al Qaeda's new brand of terrorism presented challenges to U.S. governmental institutions that they were not well-designed to meet."

The 9/11 Commission Report, National Commission on Terrorist Attacks

"Nuclear and radiological threats will persist far into the future."

Dr. Elizabeth Sherwood-Randall, Assistant to the President for Homeland Security
March 3, 2023

Are U.S. efforts to counter nuclear or radiological[1] terrorism keeping pace with the evolving threat landscape? Almost twenty years after the release of "The 9/11 Commission Report," the unanimous conclusion of the National Academies Committee members who have prepared this study report is that, overall, they are not. The nightmare scenario of a terrorist nuclear attack on U.S. soil is a risk that has not diminished. The efforts to manage this risk must be expanded and they must be enduring.

[1] The UN defines nuclear terrorism as the unlawful and intentional use of radiological material with the intent to cause death, injury, or serious damage to property or the environment, or to compel "a natural or legal person, an international organization or a State to do or refrain from doing an act" (see https://treaties.un.org/doc/db/terrorism/english-18-15.pdf). Therefore, in the rest of this document, we will use "nuclear terrorism" to refer to terrorist acts that utilize either a detonable nuclear device or radioactive substances to cause harm.

Success has the potential to breed complacency. The significant attention paid to preventing terrorism in the aftermath of the 9/11 attacks and the success of the U.S programs to reduce the terrorism risk has made it possible for a new generation of Americans to grow up without experiencing a catastrophic terrorist attack on the U.S. homeland. The downside of this achievement is that, not having had a major shock or crisis that provides confirmation that a threat is clear and present, the nation's attention has started to drift. This loss of focus, as it has in the past, can be expected in the future to translate into waning interest and investment in capabilities that are required to prevent, detect, deter, respond to, and recover from a future catastrophic terrorist event.

In the decade prior to Russia's 2022 invasion of Ukraine, most Americans paid little attention to the nuclear terrorism threat. The capture of the Zaporizhzhia Nuclear Power Station and Vladimir Putin's nuclear saber rattling placed the nuclear risk back in the headlines. This has not, however, resulted in renewed attention to nuclear terrorism. Instead, it has contributed to the accelerated shift in focus by the national security community to Great Power Competition with Russia and China.

Preventing terrorist attacks, and effectively responding to and recovering from incidents when they occur, requires vigilance and sustained effort. The sophisticated counter-terrorism intelligence and military capabilities developed after the attacks of September 11, 2001, require continuous attention to ensure the necessary levels of interagency coordination and international, state, and local engagement. But challenges presented by state actors are now being prioritized over those posed by non-state actors. While this may be understandable given the rapidly changing threat environment, it risks the erosion of efforts that have worked to date in preventing terrorist groups from obtaining or building and deploying a nuclear or radiological device.

A cautionary tale for current times is what happened to U.S. counterinsurgency capabilities in the immediate aftermath of the Vietnam War. During that conflict, the U.S. military services developed special forces that were highly capable of conducting joint special operations deep inside North Vietnam (Atlamazoglou 2020). When the war ended in 1975, however, there was a significant reduction in defense spending along with a shift by the armed services in training and strategic focus to conventional warfighting to counter the Soviet Union. Mission planning and interservice operability for conducting special operations degraded. The tragic consequence of this played out in April 1980 with the catastrophic failure of Operation Eagle Claw.

President Jimmy Carter had authorized a military rescue attempt of the 52 American diplomats and citizens taken hostage after the Iranian takeover of the U.S. Embassy in Tehran. The U.S. Army, Navy, Air Force, and Marines were all involved in the operation, but only five of eight helicopters arrived at the staging area in operational condition and then one of the remaining helicopters crashed into a transport aircraft, destroying both aircraft and killing eight servicemen. (Kyle and Eidson 2002; Williamson 2020). The operation was then aborted. The postmortem investigation concluded that "a lack of coordination between military services—evidenced in part by compartmentalized training and inadequate equipment maintenance" contributed to the aborted operation

(Lambert 2023). The failure of Operation Eagle Claw[2] illustrated how quickly "procedural memory" and defense competencies can fade once new priorities consume most of the funding and focus.

This committee is concerned that history may repeat itself, this time with respect to our counterterrorism capabilities. Fortunately, there are many dedicated people across the U.S. government who have been involved in and continue to support the successful management of the nuclear terrorism risk. These dedicated individuals recognize the imperative for sustaining capabilities and regular exercising of the capabilities needed to counter the risk.

All presidents since President Clinton have made confronting the threat of nuclear terrorism a top strategic priority. Most recently, in March 2023, President Joe Biden signed National Security Memorandum 19 (NSM-19) on "Counter Weapons of Mass Destruction Terrorism and Advance Nuclear and Radioactive Material Security" (The White House 2023).

In a meeting to coincide with the release of NSM-19, Dr. Elizabeth Sherwood-Randall, the assistant to the president for homeland security, traced the decades-long bipartisan history of confronting this risk that predates the Cold War but expanded rapidly at its end. She emphasized the Cooperative Threat Reduction (CTR) program, created by the 1991 Nunn-Lugar Act. CTR was instrumental in helping to secure and dismantle significant numbers of weapons of mass destruction and their delivery systems, secure fissile materials, and support nonproliferation programs in Russia and the states of the former Soviet Union (Bernstein and Wood 2010). Had this not been done, Soviet nuclear weapons, materials, and expertise may have ended up in the hands of rogue actors. Dr. Sherwood-Randall also discussed the challenge of addressing the decrease in high-level political attention coincident with the collective success of nuclear security. In addition, she noted that "though countering terrorism has been a top priority for the United States for more than two decades, the terrorist threat has evolved [...] It's become more ideologically diffuse, and geographically diverse" (Johnson, 2023).

This committee embraces Dr. Sherwood-Randall's admonition that "the Nation cannot lapse in this no-fail mission where the consequences are so high." Nuclear terrorism represents a uniquely consequential threat to the United States and the entire global community that is domestic, international, and transnational. This reality makes the need for sustained U.S. leadership in addressing the ongoing nuclear terrorism threat a critical national priority. This report provides a number of findings and recommendations that support existing programs but also notes where more needs to be done and where U.S leadership is indispensable. The committee members are hopeful that Congress and the American people will take these recommendations to heart and implement them. The stakes involved with getting this right could not be higher.

As mentioned previously, the committee's concerns are increasing, this time with respect to the United States' counterterrorism capabilities as events evolved while this

---

[2] Out of the Operation Eagle Claw failure came a series of congressionally led reforms that created the special operations capabilities that in 2011 conducted the successful raid (Operation Neptune Spear) into Abbottabad, Pakistan, resulting in the death of Osama bin Laden (Counterterrorism Joint Task Force 1980).

report was under review. The persistence of the terrorism threat and its capacity to be a destabilizing geopolitical force was once again highlighted by the October 7, 2023, attacks by Hamas on Israel that killed 859 Israeli civilians and at least 345 Israeli soldiers and police officers, and the taking of over 240 hostages. The attacks also demonstrated the blurred line between state and nonstate actors, given Hamas's role as a governing organization in the Gaza Strip and the support it has received from Iran and Gulf States (Boxerman 2023; Fabian 2023a, 2023b, 2023c).

Stephen E. Flynn, *Chair*
Madelyn R. Creedon, *Vice Chair*
Committee on Assessing and Improving Strategies for Preventing, Countering, and Responding to Weapons of Mass Destruction Terrorism: Nuclear Threats

References

Atlamazoglou, Stavros. 2020. "How A 'Successful Failure' Deep behind Enemy Lines 50 Years Ago Changed the Way US Special-Operations Units Plan Missions." *Business Insider*. Https://Www.Businessinsider.Com/Successful-Failure-Vietnam-War-Son-Tay-Raid-Changed-Special-Ops-2020-12.

Bernstein, Paul I., and Jason D. Wood. 2010. "The Origins of Nunn-Lugar and Cooperative Threat Reduction." https://Inss.Ndu.Edu/Portals/97/Documents/Publications/Case%20studies/Cswmd-Cs3.Pdf.

Boxerman, Aaron. 2023. "What We Know about the Death Toll in Israel from the Hamas-Led Attacks." *New York Times*, November 12, 2023. https://www.Nytimes.Com/2023/11/12/World/Middleeast/Israel-Death-Toll-Hamas-Attack.Html.

Bunn, Matthew. 2016. "The Nuclear Security Summit: Wins, Losses, and Draws." *All Nuclear Security Matters*. https://Dash.Harvard.Edu/Handle/1/37375266.

Cordesman, Anthony H. June 30, 2021. *Iran and U.S. Strategy Looking beyond The JCPOA*. Center For Strategic and International Studies (CSIS). https://www.csis.org/analysis/iran-and-us-strategy-looking-beyond-jcpoa

Counterterrorism Joint Task Force. 1980. *Holloway Commission*. Edited by Joint Chiefs of Staff.

Earnhardt, Rebecca L., Brendan Hyatt, and Nickolas Roth. 2021. "A Threat to Confront: Far-Right Extremists and Nuclear Terrorism." *Bulletin of the Atomic Scientists*. https://Thebulletin.Org/2021/01/A-Threat-To-Confront-Far-Right-Extremists-And-Nuclear-Terrorism/.

Fabian, Emanuel. 2023a. "Authorities Name 395 Soldiers, 59 Police Officers Killed in Gaza War." *The Times of Israel*, October 8, 2023. https://www.Timesofisrael.Com/Authorities-Name-44-Soldiers-30-Police-Officers-Killed-In-Hamas-Attack/.

---. 2023b. "Israeli Border Officer, 13 Palestinians Killed in West Bank Clashes, Airstrike." *The Times of Israel*, October 19, 2023. https://www.Timesofisrael.Com/10-Israeli-Officers-Hurt-5-Palestinian-Killed-In-West-Bank-Clashes-Airstrike/.

---. 2023c. "Officer Dies after Being Critically Hurt in Terror Stabbing Near Jerusalem's Old City." *The Times of Israel*, November 6, 2023. https://www.Timesofisrael.Com/2-Police-Officers-Hurt-One-Critically-In-Terror-Stabbing-Near-Jerusalems-Old-City/.

Federal Emergency Management Agency. 1996. *Emergency Management and Assistance*. Washington: U.S. Government Publishing Office.

Gill, Amandeep S. 2020. *Nuclear Security Summits: A History*. Palgrave Macmillan Cham.

Hoffman, Bruce, and Jacob Ware. 2023. "American Hatred Goes Global. How the United States Became a Leading Exporter of Whitesupremacist Terrorism." *Foreign Affairs*. Accessed September 19, 2023.

International Atomic Energy Agency. 2023. "Verification and Monitoring in the Islamic Republic of Iran in Light of United Nations Security Council Resolution 2231 (2015)." Gov/2023/24. https://www.Iaea.Org/Sites/Default/Files/23/06/Gov2023-24.Pdf.

International Panel on Fissile Materials. 2022. "Global Fissile Material Report 2022." Https://Fissilematerials.Org/Library/Gfmr22.Pdf.

Kyle, James H., and John Robert Eidson. 2002. *The Guts to Try: The Untold Story of the Iran Hostage Rescue Mission by the On-Scene Desert Commander*. Orion Crown Pub.

Lambert, Laura. 2023. "Operation Eagle Claw." *Encyclopedia Britannica*. https://www.Britannica.Com/Event/Operation-Eagle-Claw.

Lerner, K. Lee. 2022. "Policymakers Must Now Assume That Iran Has the Enriched Uranium It Needs to Build a Nuclear Weapon." *Taking Bearings. Harvard Blogs* (June 1). Https://Blogs.Harvard.Edu/Kleelerner/Iran-Now-Has-The-Enriched-Uranium-It-Needs-To-Build-A-Nuclear-Weapon/.

Murphy, Francois. 2023. "Iran Expands Stock of Near-Weapons Grade Uranium, IAEA Reports No Progress." *Reuters*, September 4, 2023. https://www.Reuters.Com/World/Middle-East/Iaea-Reports-No-Progress-Iran-Uranium-Stock-Enriched-60-Grows-2023-09-04/.

Nuclear Regulatory Commission. 1973. *10 Cfr Part 73—Physical Protection of Plants and Materials*.

The White House. 2023. *Fact Sheet: President Biden Signs National Security Memorandum to Counter Weapons of Mass Destruction Terrorism and Advance Nuclear and Radioactive Material Security*.

Williamson, Justin. 2020. *Operation Eagle Claw 1980: The Disastrous Bid to End the Iran Hostage Crisis Raid*. Osprey Publishing.

Contents

APPENDIXES

Boxes, Figures, and Tables

BOXES

FIGURES

TABLE

Acronyms and Abbreviations

AI	artificial intelligence
ARDP	Advanced Reactor Demonstration Program
CBP	Customs and Border Protection
CBRN	chemical, biological, radiological, and nuclear
CDC	Centers for Disease Control and Prevention
CFR	Code of Federal Regulations
CIA	Central Intelligence Agency
CIRP	Cesium Irradiator Replacement Project
CISA	Cybersecurity and Infrastructure Security Agency
CoM	consequence management
CONOPS	concept of operations
CPPNM	Conventional on the Physical Protection of Nuclear Materials
CRADA	Cooperative Research and Development Agreements
CRC	community reception center
CRCPD	Conference of Radiation Control Program Directors
CsCl	cesium chloride
CSI	Container Security Initiative
CSIS	Center for Strategic and International Studies
CST	civil support team
CTCP	Office of Counterterrorism and Counterproliferation
CTPAT	Customs Trade Partnership against Terrorism
CTR	Cooperative Threat Reduction
CWMD	countering weapons of mass destruction

DBT	design basis threat
DHS	Department of Homeland Security
DIA	Defense Intelligence Agency
DOD	Department of Defense
DOE	Department of Energy
DOE-NE	Department of Energy Office of Nuclear Energy
DOJ	Department of Justice
DOS	Department of State
DOT	Department of Transportation
DNN	Office of Defense Nuclear Nonproliferation
DSIC	Decision Sciences International Corporation
DTRA	Defense Threat Reduction Agency
E.O.	executive order
EPA	Environmental Protection Agency
FBI	Federal Bureau of Investigation
FEMA	Federal Emergency Management Agency
FRPCC	Federal Radiological Preparedness Coordinating Committee
FTO	Foreign Terrorist Organization
GBq	Gigabecquerel
GICNT	Global Initiative to Combat Nuclear Terrorism
GTRI	Global Threat Reduction Initiative
GW	Gigawatt
HALEU	high assay low enriched uranium
HE	high explosive
HEU	highly enriched uranium
IAEA	International Atomic Energy Agency
ICSANT	International Convention for the Suppression of Acts of Nuclear Terrorism
IMO	International Maritime Organization
IND	improvised nuclear device
INL	Idaho National Laboratory
INS	International Security Program
INSTAR	International Nuclear Security Techniques for Advanced Reactors
INTERPOL	International Criminal Police Organization
IPS	International Port Security
ISI	Islamic State of Iraq
ISIS	Islamic State of Iraq and Syria
ISPS	International Ship and Port Facility Security Code
ITWG	International Technical Working Group

JAEIC	Joint Atomic Energy Intelligence Committee
JTOT	Joint Technical Operations Team
LEU	low enriched uranium
LINAC	Linear Accelerator
LMIC	low- to middle-income countries
MDM	misinformation, disinformation, and mal-information
MOX	mixed oxide
MMPDS	Multi-Mode Passive Detection System
MPC&A	Materials Protection, Control, and Accounting
NAFTA	North American Free Trade Agreement
NARR	National Alliance for Radiological Readiness
NAS	National Academy of Science
NASEM	National Academies of Sciences, Engineering, and Medicine
NATO	North Atlantic Treaty Organization
NCBC	National Counterproliferation and Biosecurity Center
NCRP	National Council on Radiation Protection and Measurements
NCTC	National Counterterrorism Center
NDA	nondisclosure agreement
NEST	Nuclear Emergency Support Team
NGA	National Geospatial-Intelligence Agency
NHSI	Nuclear Harmonization and Standardization Initiative
NIH	National Institutes of Health
NII	nonintrusive inspection
NIC	National Intelligence Council
NIMS	National Incident Management Systems
NNSA	National Nuclear Security Administration
NOAA	National Oceanographic Atmospheric Administration
NPP	nuclear power plant
NPT	Treaty on the Non-Proliferation of Nuclear Weapons
NRC	Nuclear Regulatory Commission
NRO	National Reconnaissance Office
NSC	National Security Council
NSDD	Nuclear Smuggling Detection and Deterrence
NSM	National Security Memorandum
NSTS	National Source Tracking System
NTD	Nuclear Threat Device
O9A	Order of 9 Angles
ODNI	Office of the Director of National Intelligence
ORS	Office of Radiological Security
OSRP	Off-Site Source Recovery Project

PPE	personal protective equipment
PSI	Proliferation Security Initiative
R&D	research and development
RANET	Response Assistance Network
RDD	radiological dispersal device
RED	radiological exposure device
REMVE	racially or ethnically motivated violent extremist
REP	radiological emergency preparedness
RERT	radiological emergency response team
RIM	Russian Imperial Movement
ROSS	radiological operations support specialist
SDGT	specially designated global terrorist
SLTT	state, local, tribal, and territorial
SME	subject matter expert
SMR	small modular reactor
START	Study for Terrorism and Responses to Terrorism
TBq	Terabecquerel
TEU	twenty-foot equivalent unit
U.S.	United States
USCG	United States Coast Guard
USG	United States Government
USGS	U.S. Geological Survey
U.N.	United Nations
UNSCR 1540	U.N. Security Council Resolution 1540
WMD	weapon of mass destruction

Executive Summary

For nearly eight decades, America and the world have been navigating the dangers of the nuclear age. Despite cold war tensions and the rise of global terrorism, nuclear weapons have not been used in conflict since Hiroshima and Nagasaki in 1945.

It is no accident that the world has been spared the catastrophic consequences of a nuclear attack. Strategic deterrence, arms control, and nonproliferation agreements, U.S.-led global counterproliferation and counterterrorism efforts, and multi-year defense outlays together have helped to keep nuclear incidents at bay. A key contributor has also been decades of shared bipartisan commitment and investment in nuclear safety and security. But there is no guarantee of continued success. Indeed, there is a growing array of worrisome developments that may make nuclear terrorism more, not less, likely in the future.

The 2021 National Defense Authorization Act directed the National Academies to undertake a new study on the risks of nuclear terrorism. To respond to the broad congressional mandate, the committee conducted meetings, received briefings, and collected data from senior government officials, international partners, and experts. Drawing on the depth of expertise and experience of its dozen members, the result is the comprehensive and sobering assessment outlined in this report.

The committee does not foresee an imminent nuclear terrorist attack, but its review found that long-standing measures need fresh attention and enhancement. The committee also found gaps in programs and capabilities for interdicting nuclear weapons and materials and the nation's ability to deal with a nuclear terrorist attack.

There are several overarching findings that should guide ongoing efforts to combat nuclear terrorism and inform future budget allocations:

- Federal agencies across the U.S. government, staffed by seasoned experts, provide domestic and international programs and capabilities that have built a strong foundation for managing and reducing nuclear terrorism risks. There is, however, no agency assigned a lead role. It falls to the White House to provide active and sustained oversight to assure close interagency coordination and focus.

- Maintaining an all-of-government focus on nuclear terrorism is being challenged by the pressures on the defense and intelligence communities to shift their focus from terrorism to the risks associated with great power competition. Agencies such as the Department of Homeland Security (DHS), the Department of Energy (DOE), the Department of Defense(DOD), the Federal Bureau of Investigation (FBI), the Federal Emergency Management Agency (FEMA), the Centers for Disease Control and Prevention (CDC), the U.S. Coast Guard, and Customs and Border Protection are facing difficult choices arising from constraints on discretionary spending and the demands of their other missions. At the state and local levels, governors and mayors are confronted with competing priorities, making it difficult to devote the attention required to prepare for a low-probability/high-consequence threat. Overall, attention to the nuclear terrorism risk is waning even as important guardrails that have kept the risk in check are becoming less effective.
- Trends are pointing to a potential reversal of the post–Cold War progress in reducing the supply side of the nuclear threat. U.S.–Russian arms control agreements are lapsing with little prospect of renewal while China continues to expand its nuclear arsenal. Russia was once a major partner in advancing global nonproliferation and counterproliferation efforts, but that is no longer the case. The civil nuclear sector is expanding into countries with little experience in operating nuclear facilities and safeguarding materials.
- Terrorism is a transnational threat. U.S. domestic terrorists are developing overseas ties, and foreign terrorist organizations are recruiting Americans. This presents challenges for federal agencies that must operate under differing authorities for international versus domestic counterterrorism activities.
- Nuclear weapons, materials, and expertise are almost entirely controlled by state actors. For a terrorist organization to carry out a nuclear attack with either a nuclear weapon or an improvised nuclear device, they need either the complicity of a state or the failure of a state and its controls. In other words, the nature of nuclear terrorism is that it involves both state and non-state actors. This creates the potential for blind spots in detecting this threat if a shift in focus to state actors comes at the expense of efforts to monitor and thwart terrorist organizations.
- The radiological materials in commercial and industrial applications could be used to produce a dirty bomb or exposure device. Such a device would be less destructive than a nuclear bomb or an improvised nuclear device (IND), but the materials are easier to obtain. This makes a dirty bomb more likely to be used by terrorists than a nuclear weapon or improvised nuclear device, although the risk to lives is dramatically lower than for a nuclear weapon. But successfully exploding such a device would have significant economic consequences and

cause public fear and uncertainty. Deterring the movement of radioactive materials or the targeting of the global supply system with a dirty bomb could be strengthened by comprehensive use of advanced scanning technologies and artificial intelligence (AI).

- Under the U.S. system of government, the responsibility for disaster response and recovery lies primarily at the state and local levels where there is limited capacity for dealing with a nuclear incident. Misinformation, disinformation, and mal-information (MDM), enhanced by AI, will significantly complicate the public communications challenge for local officials in the aftermath of a nuclear attack.

The committee's 38 findings and 16 recommendations are organized around nine chapters. Chapter 1 provides an overview of the federal entities that play the most significant role in addressing the nuclear terrorism risk. Chapter 2 describes and assesses nuclear threats, and then chapter 3 looks at the evolving nature of the nuclear terrorism risk. Chapters 4 and 5 examine recent trends in the geopolitical environment and civil nuclear energy sector that are affecting this risk. Chapters 6 and 7 focus on the dangers associated with highly enriched uranium (HEU), plutonium, and radioactive source materials and the challenges of keeping them out of the hands of terrorists. Chapter 8 assesses how non-state actors might exploit well-established smuggling pathways to move nuclear weapons, materials, and equipment even in the face of current measures for detecting and interdicting such movements. Chapter 9 addresses the need for comprehensive, multi-agency emergency response and recover strategies should efforts to prevent a nuclear incident fall short. A classified annex is available as a resource with supplemental information.

Summary

"They were conquerors, and for that you want only brute force—nothing to boast of, when you have it, since your strength is just an accident arising from the weakness of others. They grabbed what they could get for the sake of what was to be got. It was just robbery with violence, aggravated murder on a great scale, and men going at it blind—as is very proper for those who tackle a darkness."

— Joseph Conrad, Heart of Darkness

This quote from Joseph Conrad's *Heart of Darkness* offers a chilling insight about the nuclear terrorism threat. A terrorist organization that gains possession of a nuclear weapon or nuclear materials would have the "strength" to take advantage of "the weakness of others"—the vulnerability of societies to a nuclear or radiological attack—resulting in "murder on a great scale."

When the Soviet Union collapsed in 1991, there was a significant risk that nuclear material and potentially even nuclear weapons might fall into the hands of non-state actors. To prevent this from happening, the United States took the extraordinary step to assist its Cold War adversary in securing nuclear weapons and weapons-usable material. This program, the Cooperative Threat Reduction (CTR) program, also known as the Nunn-Lugar program after its congressional cosponsors, former senators Sam Nunn (D-GA) and Richard Lugar (R-IN), was a bipartisan, multi-decade effort with Russia that made an important contribution to reducing the risk of nuclear and other forms of weapon of mass destruction (WMD) terrorism.

The attacks on New York and Washington by Al Qaeda on September 11, 2001, heightened the concern that weapons of mass destruction could be used by terrorist organizations. A bipartisan commission was charged by Congress to investigate the 9/11 attacks and make recommendations to bolster U.S. counterterrorism capabilities. The resulting 9/11 Commission Report helped guide efforts that have made it possible for a new generation of Americans to grow up without experiencing another catastrophic terrorist attack on the U.S. homeland, including an act of nuclear terrorism.

The nation's success to date in countering nuclear terrorism does not come with a guarantee. Success often carries the downside risk that other challenges will begin to syphon away attention and resources, and it can lead to the perception that the threat is no longer a real threat. This reality underpins Congress's decision to direct the National Academies to study whether the capabilities are still in place to keep the United States safe in the face of an evolving nuclear terrorism threat. This report is the response to that mandate and examines the status of programs and activities across the U.S. government to prevent, counter, and respond to and recover from nuclear terrorism. The report also examines state and local capacity for dealing with a nuclear incident.

A key challenge identified by the study committee is the need to ensure ongoing coordination and collaboration within and among all federal departments and agencies who bring their unique capabilities and authorities to the overall counterterrorism mission. There are strong and effective programs within a number of federal agencies staffed by experts who bring years of experience to the mission. Senior officials at the various departments and agencies work to ensure that their important domestic and international programs and capabilities are adequately funded, staffed and adapted to the evolving terrorism threat. But with no one agency assigned a lead role, it falls to the White House to provide sustained oversight to minimize duplication of efforts and to ensure close interagency coordination and focus.

As important as well-coordinated and sustained efforts are at the federal level, the committee recognizes that state and local responders will most likely be first on the scene of a nuclear terrorism event. In many instances, however, the necessary knowledge and capabilities do not exist at the state and local levels or are not exercised. Additional funding and collaboration are needed to deepen coordination among federal, state, and local government; tribal and territorial leaders; public and private universities and colleges; and other entities. New technologies and capabilities will be needed to address the evolving threat, making it essential to pursue cutting-edge research to support the mission and to grow a new generation of professionals. As in the past, philanthropy and major foundations can also play a vital role by supporting relevant work undertaken by research and policy institutes.

RECOMMENDATION: The U.S. government should maintain as a strategic priority the post-9/11 focus and effort on combating terrorism through ongoing deep collaboration and coordination across the national security community in addition to international partners; state, local, tribal, and territorial (SLTT) authorities; the national laboratories; universities and colleges; and civil society, and ensure that senior leaders at key agencies stay engaged in the counterterrorism mission.

STUDY APPROACH

The committee spent 18 months studying the means, methods, and probabilities of nuclear terrorism in response to Congress's direction to assess the adequacy of U.S. strategies to prevent, counter, and respond to nuclear terrorism. The committee was also asked to identify and provide recommendations to close any technical, policy, or resource gaps. The committee took a phased approach to the study in which it first held meetings with academic, nongovernmental, industry, and international partners. These sessions helped the committee identify the key topics on which the study should focus. These sessions guided subsequent requests for presentations from senior U.S. officials, including the review of relevant classified programs and information. The data-gathering phase was structured to understand current practices, challenges, and response needs associated with nuclear weapons, nuclear and radiological materials, and counterterrorism. The study also drew on the breadth and depth of expertise and experience of the committee's members, peer-reviewed literature, research institute reports, and investigative journalism.

The report is written to inform Congress and reach the widest possible audience, and it contains 38 findings and 16 recommendations. A classified annex contains data derived from classified sources and provides further support for the findings and recommendations found in this report.

CURRENT AND EMERGING NUCLEAR TERRORIST THREATS

Nuclear terrorist threats are generally categorized as the intentional detonation of a state-developed nuclear weapon, an improvised nuclear device (IND) assembled with stolen weapons-usable fissile material, radiological dispersal devices (RDD), radiological exposure devices (RED), or the threat to use a nuclear weapon, IND, RDD, or RED. These threats can also include attacks on nuclear facilities, including nuclear power plants.

The radiological materials in commercial and industrial applications could be used to produce a dirty bomb or exposure device. Such a device would be less destructive than a nuclear bomb or an IND, but the materials are more available and easier to obtain. Given the lower barrier for accessing these materials, a dirty bomb is more likely to be used by terrorists than a nuclear weapon or improvised nuclear device. While the risk to lives is dramatically lower than for a nuclear weapon, if an RDD is successfully exploded, it would have significant economic consequences and cause public fear and uncertainty. An RED is a more insidious weapon as it would passively expose passers-by to radiation and could go undetected for some time.

The key to preventing nuclear terrorism is to deny access to a weapon, the material, or the facility. The second line of defense is detecting and recapturing the weapons or materials, should a terrorist organization get hold of them. Since prevention efforts may not always succeed, it is also important to have in place contingencies and capabilities to deal with a nuclear or radiological device, should it be detected or used. Having the means to attribute ownership of the device will also have a deterrent effect. Countering nuclear terrorism also includes developing the means to deal with cyberattacks and managing the emerging risks associated with artificial intelligence and the use of misinformation, disinformation, and mal-information (MDM).

MDM is a particularly vexing issue that could compound the catastrophic consequences of a nuclear incident by complicating public communications, thereby impeding the emergency response and recovery efforts. MDM can be spread via many platforms, including news media or social media posts. Nation-state adversaries have used MDM to amplify extremist ideologies, to cast doubt on official narratives, amplify political discord, spark confusion, and promote favorable narratives surrounding themselves, their allies, nonaligned countries, or certain domestic actors. Given the open and widespread nature of social media, MDM could also be used by terrorists, including nuclear terrorists, to intentionally spread false information during a nuclear incident to confuse the public about what actions they should take to stay safe. Even poorly designed disinformation campaigns could impact confidence in government institutions, reputable journalistic outlets, and "other staples in democracy" (Wolters et al. 2021), making it an extremely useful tool for nuclear terrorists.

RECOMMENDATION: The Department of Homeland Security, with support from the Centers for Disease Control and Prevention, the National Governors Association, and the U.S. Conference of Mayors, should undertake a multipronged effort involving all levels of government (federal, state, local, tribal, and territorial) to include research and educational entities, civic associations, and media to raise public awareness and understanding of how information can be used to confuse, mislead, and deceive during major crises.

NEW DYNAMICS IN NUCLEAR TERRORISM POSE NEW RISKS

The committee does not foresee an imminent nuclear terrorist attack with a nuclear weapon or an IND. Nevertheless, the number and types of groups that may be motivated to use INDs, RDDs, or REDs is likely growing (Earnhardt, Hyatt, and Roth 2021). While some non-state actors may be deterred by the near-certainty of attribution and retribution, others, including millenarian groups such as ISIS and U.S.-based accelerationist groups, actively court retaliation to spark a wider war or to realize apocalyptic beliefs.

Nuclear weapons, weapons-usable fissile materials, and nuclear weapons design expertise are almost entirely controlled by state actors. This means that in order for a terrorist organization to carry out a nuclear attack with either a nuclear weapon or an improvised nuclear device, they would need the complicity of a state, the failure of the state's controls, or the failure of the state itself. As the national security and intelligence communities shift focus from terrorism to great power competition, there is danger that there will be less capability and capacity for early detection and for mobilizing a timely counterterrorism response to a non-state actor that obtains a nuclear weapon or weapons-usable fissile materials.

Importantly, terrorism is not exclusively an international or domestic threat but increasingly is a *transnational* one (Hoffman and Ware 2023). A particularly troubling development is the existence of U.S.-based accelerationist groups that have been deliberately recruiting U.S. military personnel. Additionally, there are disturbing and growing U.S. domestic links with mercenary and terrorist groups across international borders.

Another worrisome development with respect to terrorism is the extent to which technical information can be obtained online, and this could encourage groups to seek nuclear material. Additionally, extremists are utilizing social media to fuel radicalization and extreme partisanship, as well as to propagate dis- and misinformation and sow mistrust of government institutions and authoritative information. Social media is serving as a powerful organizational tool for terrorist groups, facilitating an increase in international connectivity among domestic and foreign terrorist organizations.

The risk of nuclear terrorism must also be evaluated in the context of changing norms associated with nuclear weapons and civil nuclear power. There have been cyberattacks on operating nuclear power plants in India, Japan, and South Korea. Russia has demonstrated a willingness to defy international norms, not only by attacking and occupying Ukraine's operating civilian nuclear power plants but also by employing proxies with a history of war crimes, deploying operatives to attack and poison individuals with advanced nerve agents and radiological substances, and threatening to use nuclear weapons.

In sum, managing the threat of nuclear and radiological terrorism will be challenged by the continued prevalence of groups operating both domestically and overseas who are motivated to carry out these kinds of attacks. State actors could potentially collaborate with terrorist groups, providing them the capability to conduct such attacks. As this threat landscape continues to evolve, the pressure of other national security challenges associated with great power competition along with resource constraints will make it difficult for the national defense and intelligence communities to sustain current levels of effort for managing the nuclear terrorism risk.

RECOMMENDATIONS:

- **The blurring of boundaries between state and non-state adversaries, such as the Wagner Group, Hamas, Hezbollah, and ISIS raises the possibility that there may be gaps in U.S. government efforts to address nuclear threats. The committee recommends that the National Security Council and the Office of Management and Budget conduct a review of counterterrorism programs and agency budgets across the national security community to ensure that the attention being directed to great power competition does not result in underinvesting in essential capabilities for managing and responding to the nuclear terrorism risk.**
- **The transnational links among some anti-government/terrorist groups operating in the United States suggest that some of these groups might meet the criteria to be included on the list of Foreign Terrorist Organizations, which would make it illegal not only to join these groups but also to financially support them, as is the case for other listed FTOs. The Committee recommends the Departments of State, Treasury, Justice, Defense, and other relevant agencies examine these relationships and links to under-**

stand any such international connections and determine if any additional organizations are appropriate to add to this listing.

- **To address the risk of radicalization by individuals who hold U.S. security clearances, the Administration should include the Department of Defense's revised definitions of "extremist activities" and "active participation," as described in *The Report on Countering Extremist Activity within the Department of Defense*, in the investigative standards for all government workers and contractors who have access to sensitive information and facilities as part of Executive Order 13764 of January 17, 2017, and as a part of the U.S. continuous vetting process.**

GEOPOLITICAL AND OTHER CHANGES ERODING LONG-STANDING NUCLEAR SECURITY NORMS AND PRACTICES

A unifying theme in this report is the indispensable role that the United States has played and must continue to play in mobilizing and sustaining global efforts to advance nuclear security. Renewed attention to this imperative is especially important given the erosion of many of the post–Cold War conditions that have supported international cooperation for reducing the risk of nuclear terrorism. Most prominent among these is Russia's shift from being an important partner in enhancing nuclear security to a destabilizer of nuclear norms following its full-scale invasion of Ukraine in 2022. This includes the Russian threat of using nuclear weapons to intimidate Ukraine and those countries providing assistance. A destabilizing outcome of this behavior has been the steady erosion of nonproliferation efforts by demonstrating the potential usefulness of nuclear weapons. Given this new reality, there must be continued strong U.S.-led efforts to adapt and expand the international programs that have to date prevented a successful terrorist nuclear attack and to discourage non-weapons states from acquiring nuclear weapons.

For three decades, the cornerstone of managing the nuclear terrorism threat has been limiting the number of nuclear weapons and the availability of weapons-usable nuclear materials that may potentially fall into the hands of non-state actors. In recent years, however, the global partnerships in support of arms control, nonproliferation, and combating nuclear terrorism are no longer robust.

The Nuclear Security Summit process that mobilized and focused international attention on the need to manage the risk of nuclear weapons and materials ended in 2016 (Gill 2020; Bunn 2016). Meanwhile, China and North Korea continue to expand their nuclear weapons programs, fueling the anxieties of other countries in Asia. Should the long-standing tensions between the neighboring states of Pakistan and India boil over, the fact that both states have significant and growing stocks of nuclear weapons and materials is a major concern. Iran is producing highly enriched uranium, possibly weapon usable (International Atomic Energy Agency 2023; Murphy 2023). This could stimulate interest in enrichment in other countries in the Middle East (Cordesman 2021; Lerner 2022).

Given this dynamic threat environment, there is an urgent need for the United States to reinvigorate efforts to engage heads of states and governments to work together to close any existing and emerging gaps in the international nuclear security system. Additionally, U.S. proliferation prevention programs carried out in cooperation with intergovernmental organizations like the International Atomic Energy Agency (IAEA) and Interpol, as well as with like-minded countries, require increased funding and coordination. It is clear that the current patchwork of limited bilateral and multilateral activities must be expanded, strengthened, and fully funded to manage the evolving nuclear terrorism risk.

RECOMMENDATIONS:

- **Based on the Biden Administration's recently released Strategy for Countering Weapons of Mass Destruction Terrorism (NSM-19), the U.S. government, led by the National Security Council, should continue to prioritize and provide oversight of a "whole of government" / "whole of nation" focus on preventing nuclear terrorism to include strengthening and extending ongoing nonproliferation and counterproliferation programs.**
- **Combating the threat of nuclear terrorism is a shared global interest; the U.S. government should provide strong and visible international leadership as it has done in the past.**

THE EVOLVING CIVIL NUCLEAR SECTOR: ADAPTING APPROACHES AND NEW OPPORTUNITIES

International interest in nuclear energy is growing due to its potential to provide clean power and support the goal of achieving net-zero carbon emissions. At the same time, new nuclear power technologies, including small modular reactors, are making nuclear power more accessible. Led primarily by non-U.S. corporations, the civil nuclear energy sector is now expanding into countries that lack experience with nuclear safety and safeguards. Meanwhile, Russian attacks on nuclear power plants and the civil energy sector in Ukraine have, for the first time, introduced the possibility that an operating civil nuclear power plant could be targeted by state and non-state actors as a means of coercion or terrorism.

The U.S. nuclear industry historically dominated the global market for nuclear power export throughout the 1970s and 1980s, thus collaterally exporting exceedingly high standards for nuclear safety, security, and nonproliferation. Without U.S. leadership during the upcoming wave of deployment, assurance that new entrant reactor vendors and suppliers will adopt similarly high standards may be lost. Strong U.S. leadership and presence in global markets is essential as nuclear energy technologies play a larger role in clean energy transitions around the globe. This includes forging a transparent and productive partnership among the U.S. government, the nuclear industry, and the International Atomic Energy Agency in establishing the export and adoption of high standards of safety, security, and safeguards.

To fully safeguard nuclear material, it is important to permanently dispose of spent fuel, including in the United States. An attack on a spent nuclear fuel storage could result in a radiation release although spent fuel stored in licensed storage containers, rather than in fuel pools, will be less vulnerable. Looking ahead, there will be expanding opportunities for both civilian utilities and industries to pursue nuclear power. The resulting increase in the number of civil nuclear facilities and the volume of fresh and spent fuel in transit increases the number of potential targets for terrorist attacks. More civil nuclear material and nuclear facilities around the globe will require a strategy to ensure their security from terrorist attack and proliferation for the long term.

Nuclear security is not as universally formalized and instituted as is nuclear safety. The participation by U.S. government and private sector experts in international multilateral initiatives, such as the IAEA Nuclear Harmonization and Standardization Initiative, has made a positive contribution toward achieving the goal of safe and secure deployment of small modular reactors and other advanced nuclear technologies while maximizing the potential contribution of such technologies to achieve global clean energy goals. The United States, however, needs to move beyond participation, and instead, actively lead and drive international standards-setting and regulatory harmonization efforts for attaining high standards and norms around nonproliferation, materials control and accounting, and physical and cyber security for these advanced nuclear technologies.

RECOMMENDATION: A whole-of-government effort, in partnership with the civil nuclear sector, is needed to strengthen the U.S. presence in civil nuclear energy commerce and thereby enhance global standards for safety, security, and materials control.

THE RISKS ASSOCIATED WITH HIGHLY ENRICHED URANIUM AND PLUTONIUM

Since the end of the Nuclear Security Summit process in 2016, efforts to eliminate excess civilian stockpiles of highly enriched uranium (HEU) and separated plutonium have slowed. (Together these weapons-usable fissile materials are referred to as special nuclear material.) While global inventories of HEU have remained mostly static since 2020, the inventories of plutonium have since increased by more than 17,000 kilograms, mostly as a result of commercial nuclear energy production (International Panel on Fissile Materials 2022). Five of the 31 countries with active nuclear programs—China, France, India, Japan, and Russia—use plutonium in their reactor fuel. This type of fuel cycle reprocesses spent fuel to extract the plutonium, which is the same process that a country would use to separate plutonium for nuclear weapons. While all of these countries, with the exception of Japan, already possess nuclear weapons, nuclear newcomers should be discouraged from adopting a plutonium fuel cycle that requires reprocessing. If a country does not reprocess fuel to recover plutonium, it will not have the capability and capacity to create plutonium for a nuclear weapon.

Given the evolving interest of non-state actors (terrorists, both domestic and abroad) in weapons of mass destruction, it should be a top national security priority to eliminate weapons-usable materials wherever possible and better secure those materials that are still needed. As a non-state actor does not have the ability to create these materials, it is incumbent on those 22 countries that possess these materials to make every effort to prevent them from being used by terrorists.

RECOMMENDATION: The United States should prioritize the effort to secure, and wherever practical, consolidate or eliminate civilian special nuclear materials and treat it as a core national security objective. This includes leading efforts to transform perspectives on the use of plutonium for nuclear energy production.

MANAGING THE RISKS AND BENEFITS OF RADIOACTIVE SOURCES

Radioactive sources found in commonly used tools, equipment, and critical medical devices provide many beneficial services such as cancer treatment, blood irradiation, sterilization, oil prospecting, medical research, calibration of dosimeters, food safety, and radiography. In the wrong hands, these items can be used in an RDD or RED, causing widespread panic and environmental damage. Over the past decade, Department of Energy (DOE)/NNSA (National Nuclear Security Administration) has undertaken a major effort to reduce the opportunity for terrorist use of these sources by identifying alternative technologies. These efforts include phasing out the use of high-risk cesium-137, particularly in blood irradiators, and replacing it with X-ray technology. But more attention is needed to mobilize and sustain efforts to identify additional technological alternatives, raise awareness of the risk, and enact stronger security measures. This should include working with industry and international partners to close gaps in detecting illicit source trafficking along the various pathways that terrorist groups might exploit.

Disposal costs for excess and unwanted sources can be expensive, especially for higher activity sources, and disposal facilities for these sources may not be available in many countries. In addition to known and accountable disused sources, "orphan" sources pose challenges because these sources are by definition outside of regulatory control and accounting systems and are particularly vulnerable to theft or diversion. More efforts are required to improve regulatory and accounting systems in countries across the globe to identify and eliminate orphan sources. The IAEA has guidance on how to implement effective regulatory and accounting systems. The Nuclear Regulatory Commission (NRC) via its international program office can also provide guidance to other countries, and the NRC can serve as a role model. It is also important to invest in efforts to procure and safely dispose of orphan sources.

RECOMMENDATIONS:

- **The United States, with NNSA as the lead, and in cooperation and partnership with the IAEA and other international organizations, should strengthen and accelerate current national and international activities and programs for end-of-life management of sources. Such efforts should identify disused and orphan sources and ensure that there are financial guarantees for safe and secure disposal of such materials as mentioned in a previous National Academies study (National Academies of Sciences 2021).**
- **The United States, with NNSA as the lead, and in cooperation and partnership with industry, should continue and, where feasible, expand its efforts to phase out high-risk cesium-137 and cobalt-60 sources by developing and deploying reliable alternative technologies such as X-ray irradiators. Where replacement is not feasible, the NNSA should continue to assess the security risks of facilities and develop security systems to reduce the risks attendant with cesium-137 and cobalt-60.**

DETECTION AND INTERDICTION EFFORTS WITHIN AND OUTSIDE THE GLOBAL SUPPLY SYSTEM

Non-state actors can move nuclear weapons, materials, and equipment by exploiting well-established criminal pathways for smuggling. This is true even in the face of the many detection and interdiction measures put in place since 9/11. Opportunities exist to enhance supply chain transparency and accountability by strengthening industry partnerships and taking advantage of improvements in technologies to include artificial intelligence and machine learning. Within the global supply system, these technologies can expand the means to identify anomalies and dangerous materials hidden within legitimate shipments. They can also help provide rapid forensics that can support incident response and recovery. Concurrently, strengthening efforts to counter cross-border smuggling outside the legitimate trade and travel routes also remain critical for managing the nuclear terrorism risk.

Transportation systems may not only be exploited for smuggling but potentially targeted as infrastructure critical to the economic life of the nation and global community. The COVID-19 pandemic highlighted how dependent modern economic life is on efficient cross-border supply chains, elevating the importance of strengthening the safeguards that assure the continuity of the global supply system.

RECOMMENDATIONS:

- **The United States should lead an international effort to enhance security across all elements of the global supply system by building on the post-9/11 transportation and cargo security programs and deepening international and private industry cooperation. Agencies and organizations involved**

with this effort should include the United Nations 1540 Committee, the International Maritime Organization (IMO), Department of Homeland Security (DHS) to include United States Coast Guard (USCG) and Customs and Border Protection (CBP), and Department of State (DOS), NNSA, and IAEA.

- **Department of Justice (DOJ), FBI, DOE, and DHS, with support from the U.S. DOS, should continue to deepen ongoing international law enforcement cooperation and intelligence sharing to counter nuclear smuggling efforts along illicit transit routes and between legal ports of entry. These agencies should also ensure that federal, state, local, tribal, and territorial (SLTT) law enforcement agents involved in interdiction and border control efforts receive ongoing nuclear detection training and have ready access to specialized equipment, expertise, and the means to handle radiological and nuclear materials safely.**

RESPONSE TO AND RECOVERY FROM NUCLEAR INCIDENTS

It is imprudent to assume that efforts to prevent a terrorism event will always be successful. The consequences of such an event are so catastrophic that the nation must be well prepared to respond to and recover from a nuclear incident. In the United States, there are foundations to build on that can be traced to the development of civil defense programs in the early days of the atomic age. Nuclear preparedness is almost entirely dependent on local, state, and regional authorities, most of whom are generally not adequately trained or equipped to respond to a nuclear or radiological event. Governors and mayors are confronted with competing priorities, making it difficult to devote the attention required to prepare for this kind of low-probability/high-consequence threat.

The coronavirus pandemic exposed the disparate capabilities that exist across the nation's local and state jurisdictions as well as significant shortcomings in coordination among federal, state, local, territorial, and tribal authorities in an extended public health emergency. Emergency management, like the U.S. public health system, operates primarily under the purview of governors, county commissioners, and mayors for which the federal government plays a support role. In a nuclear incident, consequence management and recovery personnel have the added burden of managing it in the face of widespread fear. The complexity will increase if inaccurate information is widely disseminated, either intentionally or unintentionally. An adequate response to a nuclear or radiological incident requires enhanced coordination of emergency management response protocols across all levels of government and protocols and experts to provide accurate information and disseminate trusted, science-based information.

Significant new investments in resources would likely be needed to develop and sustain adequate nuclear incident response and recovery capabilities at the local and state levels.

RECOMMENDATIONS:

- **Federal Emergency Management Agency (FEMA) should reinvigorate a dynamic, comprehensive, and inclusive exercise regimen, in coordination with the Federal Radiological Preparedness Coordinating Committee (FRPCC) and with guidance and oversight from the NSM-19-established council of leadership. This should include fully utilizing the FRPCC (Federal Emergency Management Agency 1996; Nuclear Regulatory Commission 1973) in its capacity as a national-level forum to develop and coordinate radiological prevention and preparedness policies and procedures.**
- **FEMA with Centers for Disease Control and Prevention (CDC), Environmental Protection Agency (EPA), DOE, and National Institutes of Health (NIH) should empower local response by making available simple and accessible real-time information through application development that will facilitate standardized actions and guide an appropriate public response. To assist, the White House should clarify the agency that serves as overall lead for providing federal interagency coordination and oversight of developing response tools to include educating state, local, tribal, and territorial officials as well as the general public on their availability and utility, and it should strive to establish itself as a trusted agent.**
- **The president should request and Congress should support adequate resources for consequence management (CoM) programs that are key to a nuclear incident response. This should recognize the important role states, localities, tribal nations, and territories play in saving lives. More resources are needed because these programs have insufficient budgets, staffs, and capabilities, and yet are foundational to any successful response to a nuclear or radiological event.**

References

Bunn, M. 2016. "The Nuclear Security Summit: Wins, Losses, and Draws." *All Nuclear Security Matters*. https://dash.harvard.edu/handle/1/37375266.

Cordesman, AH. June 30, 2021. *Iran and U.S. Strategy Looking Beyond the JCPOA*. Center for Strategic and International Studies (CSIS). http://www.jstor.org/stable/resrep33760.

Earnhardt, RL., B. Hyatt, and N. Roth. 2021. "A Threat to Confront: Far-right Extremists and Nuclear Terrorism." *Bulletin of the Atomic Scientists*. https://thebulletin.org/2021/01/a-threat-to-confront-far-right-extremists-and-nuclear-terrorism/.

Gill, AS. 2020. *Nuclear Security Summits A History*. Palgrave Macmillan Cham.

Hoffman, B., and J. Ware. 2023. "American Hatred Goes Global. How the United States Became a Leading Exporter of WhiteSupremacist Terrorism." *Foreign Affairs*. Accessed September 19, 2023.

International Atomic Energy Agency. 2023. "Verification and Monitoring in the Islamic Republic of Iran in Light of United Nations Security Council Resolution 2231 (2015) " GOV/2023/24. https://www.iaea.org/sites/default/files/23/06/gov2023-24.pdf.

International Panel on Fissile Materials. 2022. "Global Fissile Material Report 2022." https://fissilematerials.org/library/gfmr22.pdf.

Lerner, KL. 2022. "Policymakers Must Now Assume That Iran Has the Enriched Uranium It Needs to Build a Nuclear Weapon." *Taking Bearings. Harvard Blogs* (June 1). https://blogs.harvard.edu/kleelerner/iran-now-has-the-enriched-uranium-it-needs-to-build-a-nuclear-weapon/.

Murphy, F. 2023. "Iran Expands Stock of Near-weapons Grade Uranium, IAEA Reports No Progress." *Reuters*, September 4, 2023, 2023. https://www.reuters.com/world/middle-east/iaea-reports-no-progress-iran-uranium-stock-enriched-60-grows-2023-09-04/.

National Academies of Sciences, Engineering, Medicine,. 2021. *Radioactive Sources: Applications and Alternative Technologies*. Washington, DC: The National Academies Press.

Wolters, H., K. Stricklin, N. Carey, and M. McBride. 2021. *The Psychology of (Dis) information: A Primer on Key Psychological Mechamisms*. Center for Naval Analyses. https://www.cna.org/reports/2021/10/The%20Psychology-of-%28Dis%29information-A-Primer-on-Key-Psychological-Mechanisms.pdf.

FIGURE 1-1 A member of the Russian military guarding the Zaporizhzhia nuclear power plant in southeastern Ukraine, May 2022. The capture of the Zaporizhzhia nuclear power plant and Vladimir Putin's nuclear saber rattling during the course of this consensus study highlight that increased volatility in the global security environment can exacerbate nuclear risks.
SOURCE: AP, https://www.npr.org/2023/07/05/1185980734/zelenskyy-warns-sabotage-nuclear-plant.

1

Background and Study Task

1.1 CHARGE TO THE COMMITTEE

Spurred by an increasingly dynamic international security environment, Congress included a mandate in the National Defense Authorization Act for Fiscal Year 2021 (Section 1299I. Assessment of Weapons of Mass Destruction Terrorism [U.S. Congress 2021]) that the National Academies of Sciences, Engineering, and Medicine conduct an assessment of U.S. strategies for preventing, countering, and responding to nuclear terrorism, and make recommendations to improve such strategies. The committee tasking came prior to Russia's February 2022 invasion of Ukraine and is included in Box 1-1.

In response to this congressional mandate, the National Academies assembled a committee of experts (referred to as "the committee" in this report) to assess the current strategies and nuclear risk, identify potential gaps in policy, and examine new approaches to the current challenges. The committee consisted of 12 volunteer members and an unpaid consultant, all with years of experience and a broad range of careers spanning (1) the U.S. government including the Department of Energy (DOE), the Department of Defense (DOD), the National Nuclear Security Administration (NNSA), the Department of Homeland Security (DHS), the National Security Council, and the Nuclear Regulatory Commission (NRC); (2) the U.S. Armed Services; (3) NNSA National Laboratories; (4) academia; and (5) nongovernmental organizations.

As directed in the statutory language, this report provides suggested actions for Congress, the White House, and relevant federal agencies to enhance the security of the United States and its allies and partners in managing the nuclear terrorism risk. The report is also written to contribute directly to the general public's understanding of this ongoing risk, embracing a recognition by the committee that an informed and engaged civil society is critical to preventing, responding to, and recovering from a nuclear terrorist attack.

BOX 1-1
Statement of Task

The National Academies of Sciences, Engineering, and Medicine (NASEM) will appoint an ad hoc topical committee to address specific issues related to nuclear terrorism threats. This committee will address the adequacy of strategies to prevent, counter, and respond to nuclear terrorism, and identify technical, policy, and resource gaps with respect to:

1. identifying national and international nuclear risks, and critical emerging threats;
2. preventing state-sponsored and non-state actors from acquiring or misusing the technologies, materials, and critical expertise needed to carry out nuclear attacks, including dual-use technologies, materials, and expertise;
3. countering efforts by state-sponsored and non-state actors to carry out such attacks;
4. responding to nuclear terrorism incidents to attribute their origin and help manage their consequences;
5. budgets likely to be required to implement effectively such strategies; and
6. other important matters that are directly relevant to such strategies.

The NASEM will produce a consensus report and may produce additional products (such as proceedings of workshops) by mutual agreement with the sponsor. The consensus report will be unclassified with a classified annex.

1.2 STUDY APPROACH

Over the course of 18 months, the committee took a phased approach, building a foundation for the study. Data gathering focused on briefings and panel discussions from experts in the relevant areas and from entities focused on the risks associated with nuclear weapons of mass destruction. The study process incorporated the expertise and experience of the committee, peer-reviewed literature, press reports, and classified reports. The committee received briefings from U.S. government agencies and outside experts to include classified briefings (detailed agendas can be found in Appendix A). This phase also included briefings and discussion with key international partners. Specifically, the committee heard from the following entities:

- academia: Global Resilience Institute at Northeastern University, Belfer Center for Science and International Affairs at Harvard University, Program on Crisis Leadership at Harvard University, Frederick S. Pardee School of Global Studies at Boston University, Oregon State University, and the University of California;

- nongovernmental organizations: Nuclear Threat Initiative, Ploughshares Fund, The Stimson Center, and The Atlantic Council;
- journalists: the *National Journal*, *The Atlantic*, and the *New York Times*; and
- U.S. international partners and allies: Embassy of Canada to the United States, United Nations, INTERPOL, NATO, and the Canadian Nuclear Safety Commission.

This unclassified report contains the conclusions, findings, and recommendations from the study committee that appear in chapters 2 through 9 and are compiled in the report summary. The report also has a classified annex that contains additional information pertaining to sensitive information and restricted programs managed by U.S. government departments and agencies.

1.3 THE U.S. GOVERNMENT AGENCIES FOCUSED ON NUCLEAR TERRORISM

Federal agencies across the U.S. government, staffed by seasoned experts, provide domestic and international programs and capabilities that have built a strong foundation for managing and reducing nuclear terrorism risks. The committee received reports and presentations and had panel discussions with nearly all the key U.S. government agencies involved in this mission. These are captured in Figure 1-2 and described below.

Many of the relevant programs for dealing with the enduring and evolving nuclear threat are funded and managed by the U.S. DOE's NNSA. The NNSA defines its role along three lines of effort (National Nuclear Security Administration 2021b):

- "**Prevent** proliferant states from developing nuclear weapons or acquiring weapons-usable nuclear materials, equipment, technology, and expertise, and prevent non-state actors from acquiring nuclear and radioactive materials that can be used for malicious purposes;
- "**Counter** the efforts of both proliferant states and non-state actors to acquire, develop, disseminate, deliver, or use the materials, expertise, or components of a nuclear or radiological device; and
- "**Respond** to the full spectrum of nuclear and radiological emergencies at home or abroad, including deliberate attacks and accidents, to minimize the damage from such incidents."

These lines of effort are the focus of three offices at NNSA that report directly to the administrator: (1) The Office of Defense Nuclear Nonproliferation (DNN) develops and implements policy, programmatic, and technical solutions to eliminate or secure fissile and sensitive radiological materials and limit or prevent the proliferation of materials, technology, and expertise related to nuclear and radiological weapons, interfacing with other federal agencies (National Nuclear Security Administration 2021a). (2) The Office of Counterterrorism and Counterproliferation (CTCP) works closely with the DOD and the Federal Bureau of Investigation (FBI) to counter nuclear threats and respond to

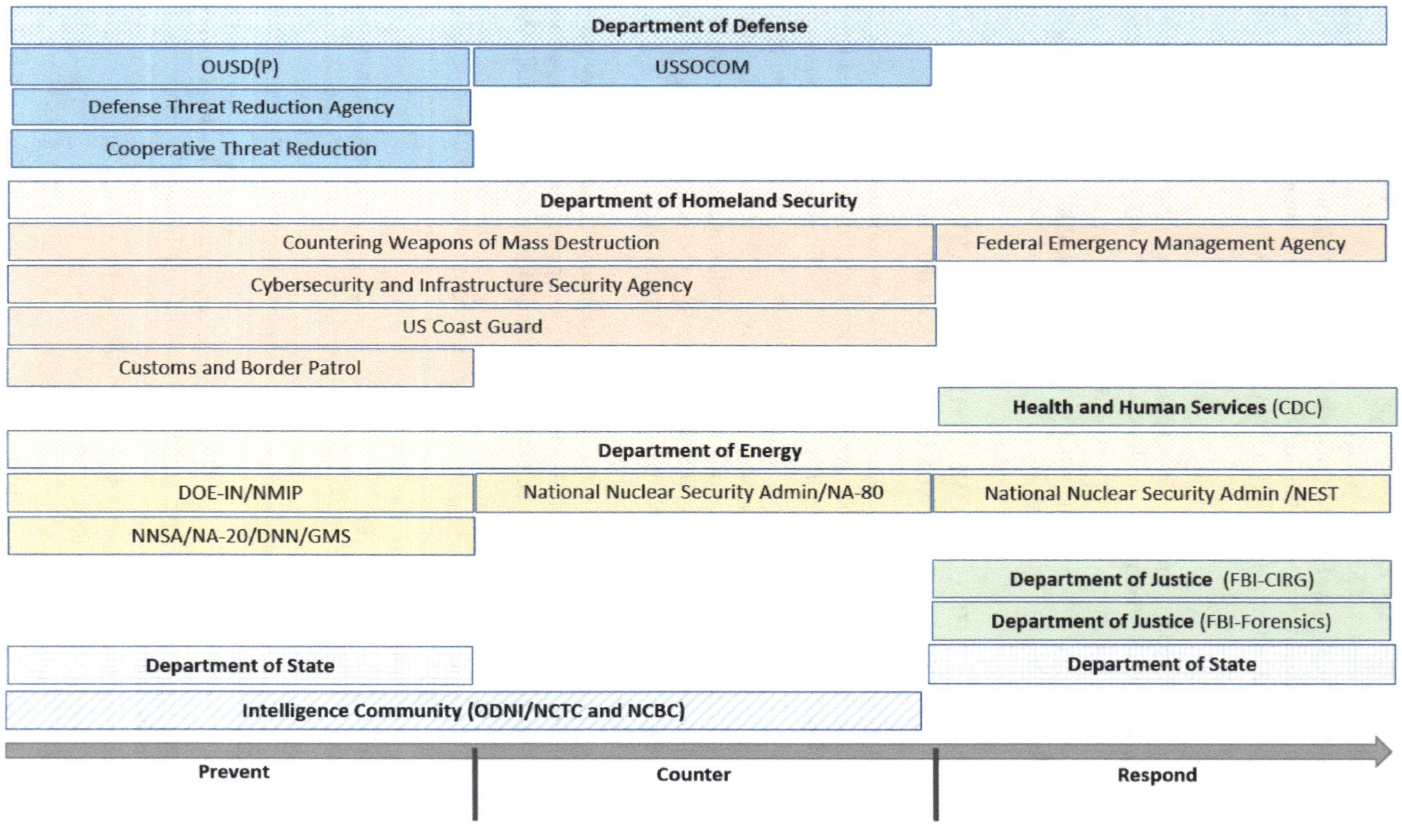

FIGURE 1-2 An abridged list of U.S. government agencies focused on nuclear terrorism that were able to brief the Academies committee, mapped to the missions of preventing, countering, and responding to nuclear threats.

nuclear incidents and accidents, domestically and abroad. (3) The Office of Emergency Operations administers and directs the implementation and integration of emergency management programs across the DOE, including the NNSA.

The 17 national laboratories in the DOE/NNSA complex support these missions for NNSA through research, development, testing, and implementation. These include the NNSA-managed laboratories: Los Alamos National Laboratory, Sandia National Laboratories, and Lawrence Livermore National Laboratory. Other DOE laboratories—including Oak Ridge National Laboratory (along with Y-12), Pacific Northwest National Laboratory, Savannah River National Laboratory, Idaho National Laboratory, Argonne National Laboratory, and Brookhaven National Laboratory—provide considerable support to the counterterrorism mission, domestically and internationally, with other programs scattered throughout the remaining national laboratories (Ashby 2021; Center for Global Security Research 2023; Nicholas 2016; Nicholas and Smith 2022; Pepper et al. 2016; Rowland 2020; Savannah River National Laboratory 2022; Shirey 2018).

The NNSA funds and manages a variety of programs that address international threats and risks, but there are several federal departments and agencies that play important roles in preventing, countering, and responding to nuclear terrorism. The Department of State (DOS), DOD, DHS, FBI, and NRC all fund and manage programs for preventing and responding to nuclear terrorism domestically.

The DHS—formed after the 9/11 terrorist attacks along with its key components to include the U.S. Coast Guard, Customs and Border Protection, FEMA, TSA, and the Countering Weapons of Mass Destruction Office—coordinates efforts to prevent, respond to, and recover from nuclear terrorism incidents and enhance security measures at ports and borders, and it works to detect and prevent the illicit trafficking of nuclear materials. The FBI plays a critical role in national counterterrorism efforts, including countering nuclear terrorism. The Bureau investigates threats, gathers intelligence, and works to disrupt and dismantle networks involved in nuclear terrorism. In addition to intelligence activities, the Critical Incident Response Group and the forensics team within the FBI have key roles for managing the investigation in the aftermath of a nuclear incident.[1] The NRC regulates the civilian use of nuclear materials in the United States, preventing nuclear terrorism by preventing the proliferation of nuclear and radiological materials at nuclear power plants, fuel cycle facilities, and other licensed nuclear facilities. The NRC sets regulations and standards for physical protection measures, safeguards, and emergency preparedness to mitigate the risks of nuclear terrorism at nuclear facilities.

The DOD has specialized activities for managing the nuclear terrorism risk to include pathway defeat; weapon of mass destruction (WMD) defeat; and chemical, biological, radiological, and nuclear (CBRN) response (U.S. Joint Chiefs of Staff 2019). Listed below are the relevant DOD lines of effort:

[1] Written materials submitted to a study committee by external sources and public meeting recordings are listed in the project's Public Access File and can be made available to the public upon request. Contact the Public Access Records Office (PARO) at the National Academies of Sciences, Engineering, and Medicine for a copy of the list and to obtain copies of the materials. E-mail: paro@nas.edu.

- threat assessment and intelligence sharing, contributing to the overall understanding of nuclear terrorism risks;
- deterrence and defense, maintaining a strong nuclear deterrent posture to dissuade potential adversaries from using nuclear weapons or supporting nuclear terrorism;
- nuclear nonproliferation and arms control;
- building capability with international partners to counter and deter weapons of mass destruction and emerging threats; and
- crisis response and consequence management, working alongside other federal, state, and local, tribal, and territorial agencies to provide support, including emergency response coordination, intelligence analysis, logistics, medical assistance, and technical expertise.

The DOS leads diplomatic efforts and supports capacity-building for preventing nuclear proliferation and nuclear terrorism. These diplomatic efforts include engaging international partners in actively participating in nonproliferation initiatives and abiding by legal frameworks for preventing, countering, and responding to nuclear terrorism threats. The DOS plays an essential role in building international consensus, promoting security cooperation, strengthening international organizations and treaties, and addressing the global challenges posed by nuclear terrorism. The DOS is also involved with Cooperative Threat Reduction (CTR) programs to include efforts by the Bureau of International Security and Nonproliferation in building the capacity for dealing with insider threats and bolstering security culture.

Within the intelligence community (IC), a number of agencies to include the CIA, Defense Intelligence Agency (DIA), National Geospatial-Intelligence Agency (NGA), National Reconnaissance Office (NRO), and NSA play important roles in this mission, as do important entities for advancing interagency coordination within ODNI. These include the National Counterterrorism Center (NCTC), the National Counterproliferation and Biosecurity Center (NCBC), the National Intelligence Council (NIC), and the Joint Atomic Energy Intelligence Committee (JAEIC) (National Intelligence Council Joint Atomic Energy Intelligence Committee 2023; Office of the Historian U.S. Department of State 2023).

The NCTC mission is to "lead and integrate the counterintelligence effort by fusing all U.S. government counterintelligence information, providing terrorism analysis, sharing information with partners across the counterintelligence enterprise, and driving whole-of-government action to secure our national counterintelligence objectives" (National Counterterrorism Center 2021).

The NCBC advances the broader objectives of the U.S. counterproliferation mission by discouraging interest by states, terrorists, or armed groups in acquiring, developing, or mobilizing resources for WMD purposes and preventing or obstructing state, terrorist, or other efforts to acquire WMD capabilities, or efforts by suppliers to provide such capabilities (National Counterproliferation and Biosecurity Center 2023). The NCBC is responsible for integrating the intelligence community's collection and analytic efforts against the highest-priority threats, including atomic energy and nuclear weapons development in foreign countries.

The primary role of the NIC is to provide strategic intelligence assessments to senior policymakers, including the president, the National Security Council, and other key decision-makers (National Intelligence Council 2023). The NIC's role is crucial in helping policymakers understand the complex global landscape, identify potential threats and opportunities, and make informed decisions that shape U.S. national security and foreign policy. This includes collaborating with nongovernmental organizations in academia and the private sector.

Finally, the JAEIC has the responsibility to assess foreign atomic energy developments and contribute to national intelligence products, keeping the ODNI aware of foreign atomic energy intelligence regarding national security concerns (National Intelligence Council Joint Atomic Energy Intelligence Committee 2023).

Given the number of federal departments and agencies and the depth and breadth of the unique capabilities and authorities that they bring to the counterterrorism mission, a central challenge that the committee identified is making sure there is all-of-government focus on the mission. Senior leaders at the relevant departments and agencies should prioritize this mission, and the White House will have to provide ongoing oversight that ensures close interagency coordination and focus.

1.4 ORGANIZATION OF THE REPORT

The report is organized into the following chapters:

Chapter 2	Nuclear Terrorism Threats
Chapter 3	New Dynamics in Nuclear Terrorism Pose New Risks
Chapter 4	Geopolitical and Other Changes Eroding Long-Standing Nuclear Security Norms and Practices
Chapter 5	The Evolving Civil Nuclear Energy Sector: Adapting Approaches and New Opportunities
Chapter 6	The Risks Associated with Highly Enriched Uranium and Plutonium
Chapter 7	Managing the Risks and Benefits of Radioactive Sources
Chapter 8	Detection and Interdiction Efforts within and outside the Global Supply System
Chapter 9	Response to and Recovery from Nuclear Incidents

Chapter 2 describes and assesses the tools and types of nuclear terrorism. Chapter 3 looks at the evolving and increasingly transnational nature of the nuclear terrorism risk. Chapters 4 and 5 examine recent trends in the geopolitical environment and civil nuclear energy sector that are affecting this risk. Chapters 6 and 7 focus on the dangers and challenges of keeping highly enriched uranium (HEU), plutonium, and radioactive source materials out of the hands of terrorists. Chapter 8 assesses how non-state actors might exploit well-established smuggling pathways to move nuclear weapons, materials, and equipment even in the face of current measures for detecting and interdicting such movements. Chapter 9 addresses the need to plan for and be prepared to respond and recover from a nuclear incident should prevention efforts fall short. The sensitive details and programs specific to efforts to prevent and counter nuclear terrorism are compiled in the classified annex. Note that included in this report is an appendix capturing the recent eras in nuclear security that is in response to the committee's mandate for this study (Appendix B). The committee felt it was important and valuable for stakeholders and decision-makers to put in context the historical eras for nuclear security.

1.5 PREVIOUS/RELEVANT NATIONAL ACADEMIES REPORTS

The National Academies has published many reports relevant to this study to include nuclear risk assessment, international nuclear policy, and nuclear/radioactive threat reduction (National Academies of Sciences 2023). The committee looked to these previous studies to help inform this report and build on these comprehensive prior efforts.

1. **Nuclear Risk and Threat Reduction:** There are many factors that influence the probability and severity of a nuclear terrorist attack. Previous National Academies activities have gone into detail on specific factors such as violent extremism, government workforce requirements, and nuclear material management. In-depth analysis of all these factors falls outside of the scope of this committee. However, this report builds on these earlier findings and relevant conclusions and recommendations contained in:
    - *Governance and Management of the Nuclear Security Enterprise* (2020)
    - *Scientific Aspects of Violent Extremism, Terrorism, and Radiological Security Proceedings of a Workshop—in Brief* (2020)
    - *Cooperative Threat Reduction Programs for the Next Ten Years and Beyond: Proceedings of a Symposium—in Brief* (2018)
    - *Emerging and Readily Available Technologies and National Security: A Framework for Addressing Ethical, Legal, and Societal Issues* (2014)
    - *Assuring a Future U.S.-Based Nuclear and Radiochemistry Expertise* (2012)
    - *Global Security Engagement: A New Model for Cooperative Threat Reduction* (2009)
    - *Making the Nation Safer: The Role of Science and Technology in Countering Terrorism* (2002)
2. **International Policy and Collaboration:** The United States and Russia together possess the vast majority of the world's nuclear weapons and for nearly three decades shared the largest responsibility for mitigating nuclear risk (Arms Control Association 2023; Federation of American Scientists 2023). International cooperation from multiple parties is imperative to prevent proliferation across international

borders and ensure the security of all countries. The National Academies has a long history of engaging with Russia and other key international stakeholders. The committee's work was informed by these historic NAS exchanges with Russia and other international partners along with the lessons learned and the recommendations derived from programs to support international nuclear policy formulation. These include:

- *Roots and Trajectories of Violent Extremism and Terrorism: A Cooperative Program of the U.S. National Academy of Sciences and the Russian Academy of Sciences* (1995–2020; 2022)
- *Improving International Resilience and Response to Chemical, Biological, Radiological, and Nuclear Events* (2019)
- *Brazil–U.S. Workshop on Strengthening the Culture of Nuclear Safety and Security* (2015)
- *Future of the Nuclear Security Environment in 2015: Proceedings of a Russian-U.S. Workshop* (2009)
- *Internationalization of the Nuclear Fuel Cycle: Goals, Strategies, and Challenges* (2009)
- *Russian Views on Countering Terrorism During Eight Years of Dialogue: Extracts from Proceedings of Four U.S.-Russian Workshops* (2009)
- *U.S.-Russian Collaboration in Combating Radiological Terrorism* (2007)
- *Strengthening U.S.-Russian Cooperation on Nuclear Nonproliferation* (2005)
- *Overcoming Impediments to U.S.-Russian Cooperation on Nuclear Nonproliferation- Report of a Joint Workshop* (2004)
- *Scientists, Engineers, and Track-Two Diplomacy—A Half-Century of U.S.-Russian Interacademy Cooperation* (2004)

3. **Nuclear Forensics and Technology:** Increased interest and funding to support the development of civil nuclear technology has led to recent innovation in nuclear reactors and material detection. The National Academies studies listed below have assessed the landscape and implications of new technology; this report draws from and expands on these reports.
   - *Merits and Viability of Different Nuclear Fuel Cycles and Technology Options and the Waste Aspects of Advanced Nuclear Reactors* (2023)
   - *Understanding the Societal Challenges Facing Nuclear Power: Proceedings of a Workshop* (2022)
   - *Radioactive Sources: Applications and Alternative Technologies* (2021)
   - *Nuclear Proliferation and Arms Control Monitoring, Detection, and Verification: A National Security Priority: Interim Report* (2021)
   - *Restoring and Improving Nuclear Forensics to Support Attribution and Deterrence: Public Summary* (2021)
   - *Monitoring Nuclear Weapons and Nuclear-Explosive Materials: An Assessment of Methods and Capabilities* (2005)

References

Arms Control Association. 2023. "Nuclear Weapons: Who Has What at a Glance." Last Modified June 2023. https://www.Armscontrol.Org/Factsheets/NuclearweaponsWhohaswhat.

Ashby, Steven. 2021. "Looking Forward from 9/11: PNNL Contributes to Homeland Security." *Tri-City Herald*, September 20, 2021, 2021. https://www.Pnnl.Gov/News-Media/Looking-Forward-911-Pnnl-Contributes-Homeland-Security.

Center for Global Security Research. 2023. Lawrence Livermore National Laboratory https://Cgsr.Llnl.Gov/About.

Federation of American Scientists. 2023. "Nuclear Weapons: Nuclear Information Project." https://Fas.Org/Issue/Nuclear-Weapons/.

National Academies of Sciences, Engineering, and Medicine,. 2023. "Aiding Global Security through Research." https://Sites.Nationalacademies.Org/Pga/Pga_041596.

National Counterproliferation and Biosecurity Center. 2023. "What We Do." https://www.Dni.Gov/Index.Php/Ncbc-What-We-Do.

National Counterterrorism Center. 2021. Inside NCTC.

National Intelligence Council. 2023. "National Intelligence Council—What We Do." https://www.Dni.Gov/Index.Php/Who-We-Are/Organizations/Mission-Integration/Nic/Nic-What-We-Do

National Intelligence Council Joint Atomic Energy Intelligence Committee. 2023. *Subject: Charter for the Joint Atomic Energy Intelligence Committee, Annex to Intelligence Community Directive Number 207.* https://www.Dni.Gov/Files/Documents/Foia/Df-2017-00410-Ravnitzky-Final-Response.Pdf

National Nuclear Security Administration. 2021a. NNSA: Reducing Global Nuclear Threats.

---. 2021b. "Prevent, Counter, and Respond—A Strategic Plan to Reduce Global Nuclear Threats (NPCR)." https://www.Energy.Gov/Nnsa/Articles/Prevent-Counter-And-Respond-Strategic-Plan-Reduce-Global-Nuclear-Threats-Npcr.

Nicholas, Nancy Jo. 2016. "Los Alamos National Laboratory's Role in Implementing U.S. Nuclear Policy." Conference: NNSS International Diplomat Visit; 2016-10-25 - 2016-10-25; Las Vegas, NV. https://www.Osti.Gov/Biblio/1331250.

Nicholas, Nancy Jo, and Jim Smith. June 29, 2022. *Programmatic Capability Awareness Workshop: Global Security.* Los Alamos National Laboratory. https://Permalink.Lanl.Gov/Object/Tr?What=Info:Lanl-Repo/Lareport/La-Ur-22-26248.

Office of The Historian U.S. Department of State. 2023. "126. Director of Central Intelligence Directive No. 3/4." https://History.State.Gov/Historicaldocuments/Frus1950-55intel/D126.

Pepper, S., T. Mckinney, and L. Satkowiak. 2016. "U.S. National Laboratory Contributions to Global Nuclear Security." Conference: International Conference on Nuclear Security: Commitments and Actions; Vienna, Austria; 20161205 through 20161205, United States. https://www.Osti.Gov/Biblio/1338603.

Rowland, Lynzie Rae. 2020. *Global Security: 5–10 Year Strategy [Slides].* Sandia National Laboratories (United States). https://www.Osti.Gov/Biblio/1598759.

Savannah River National Laboratory. 2022. "SRNL Global Security Directorate." https://Srnl.Doe.Gov/Factsheets_2023/Gsd.Pdf.

Shirey, Robert. 2018. *Nuclear Nonproliferation and Security Program Office–Highlights 2017.* https://www.Osti.Gov/Biblio/1430026.

U.S. Congress. 2021. William M. (Mac) Thornberry National Defense Authorization Act for Fiscal Year 2021.

U.S. Joint Chiefs of Staff. 2019. Joint Countering Weapons of Mass Destruction. Edited By Department of Defense.

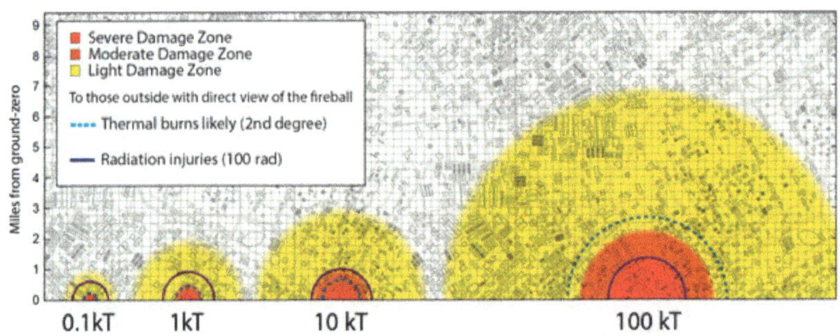

FIGURE 2-1 Radiation and burn injury ranges overlaid on damage zones demonstrating the extent of outdoor 1 Gy (100 rad) initial radiation and second-degree thermal burns for unobstructed 0.1, 1, 10, and 100 kT surface detonations. These data depict examples of the potential devastating damage from nuclear weapons but does not capture what would also be significant psychological damage nationwide. Also discussed in this chapter are other threats, including improvised nuclear devices, radiological dispersal and exposure devices, and physical and cyberattacks on nuclear power plants.
SOURCE: Federal Emergency Management Agency 2022.

2

Nuclear Terrorism Threats

2.1 INTRODUCTION TO NUCLEAR THREATS

In 2008, while the terrorist attacks of 9/11 were still fresh in the minds of Americans, the Commission on the Prevention of Weapons of Mass Destruction Proliferation and Terrorism submitted its final report to President George W. Bush and the U.S. Congress (Graham et al. 2008). This Commission was tasked to "assess ... any, and all of the nation's activities, initiatives, and programs to prevent weapons of mass destruction proliferation and terrorism" (U.S. Congress 2007). The resulting report, titled "World at Risk," identified important, actionable recommendations to address the Weapons of Mass Destruction (WMD) threat.

The WMD Commission focused on the perilous crossroads of terrorism and proliferation in poorly governed parts of the world, and the prevention of biological and nuclear terrorism. It also analyzed the potential erosion of international nuclear security, treaties, and norms as the world entered a period of expected growth in nuclear energy production. Given the many recent destabilizing geopolitical developments, all these issues remain relevant today and reinforce the germaneness of the Commission's call for vigilance in addressing the ongoing nuclear terrorism risk.

Fifteen years have passed since the WMD Commission's report was completed, and there has been no known terrorist acquisition of a nuclear weapon, improvised nuclear device, or radiological dispersal device. Still, Al Qaeda showed interest in nuclear terrorism to include conducting inert-material implosion testing in Afghanistan, undertaking efforts to steal materials and recruit scientists (Albright 2010; Mowatt-Larssen 2010, 2020). On August 4, 2014, the Doel 4 nuclear power plant in Belgium was shut down automatically as a result of an act of sabotage by an unidentified organization. Notwithstanding the many changes in the international security environment, the risk of nuclear terrorism remains significant.

With the demise of the Soviet Union, the U.S. government instituted very effective cooperative programs with the Russian Federation to dismantle Soviet weapons and

BOX 2-1
Summary

The world faces an ongoing risk that non-state actors will gain access to and use a nuclear weapon. A more likely threat is that a terrorist organization will seek out fissile materials and designs to assemble an improvised nuclear device (IND), a radiological dispersal devise (RDD), or radiological exposure device (RED). There is also the potential for a physical or cyberattack on nuclear facilities that could include power plants. The current information age is generating new capabilities for terrorists to obtain sensitive nuclear-related information. It is also providing the means to create confusion in the aftermath of a nuclear incident by using misinformation, disinformation, and mal-information (MDM). Preventing, countering, and responding to nuclear terrorism focuses on denying access to nuclear and radiological material, recapturing illegally acquired material, and exercising the tools necessary to respond to the discovery or use of a nuclear device. These efforts have not been fully adapted to the changing nature of terrorism and to dangers associated with artificial intelligence (AI) and other information technologies.

Highlights

- Weapons-usable nuclear material remains at risk. While obtaining state weapons remains difficult for non-state actors, there are sources of nuclear and radiological materials that are not as secure. Also, nuclear power plants can be targeted by state and non-state actors that could lead to significant radiation dispersion.
- Construction of an improvised nuclear device with fissile material is challenging. Radiological dispersal devices (dirty bombs) and radiological exposure devices, which have designs that are more simple than nuclear weapons, remain persistent concerns.
- Sabotage at nuclear facilities by knowledgeable insiders has occurred. Concern about cyberattacks is growing with the targeting of nuclear facilities becoming more commonplace, varying in the degree of sophistication and threat, by both state and non-state actors.
- Information operations by nuclear terrorists that tap into the widespread dread of radiation and nuclear weapons could be very influential and harmful given growing public susceptibility to MDM.
- New developments and access to AI could make MDM even more damaging with unknown consequences and generate added challenges to preventing and countering influence operations by terrorists.

weapon infrastructure and improve the security of components and materials of concern. Programs (among them U.S.-former Soviet Union (FSU) science cooperation, the DOE Nuclear Cities Initiative, the U.S. Civilian Research and Development Foundation, and the "Lab to Lab" program) also were established to engage Soviet weapon scientists in non-weapon work, with the intent of reducing the risk that their skills might be procured by nefarious entities (National Research Council 1996; Rotblatt 1998). In addition, the United States and Russia completed the "Megatons to Megawatts" program, which eliminated 500 metric tons of highly enriched uranium (HEU) by blending it down to low-enriched uranium fuel for civilian nuclear reactors.

Much has changed in the strategic environment since these successful programs were put in place, especially with respect to Russia's relationship with the United States. Today there are other nuclear-armed states that could experience instability and governance challenges that would potentially result in a loss of control of nuclear weapons, fissile materials, or expertise. Possible examples include economic collapse in Pakistan or the fall of the Kim regime in North Korea. As the breakup of the Soviet Union demonstrated, such events can be sudden and difficult to predict.

Among the disturbing changes to the global security landscape is that terrorist organizations have proliferated, with a growing number of terrorists' movements operating globally and transnationally. Political polarization within the United States is also on the rise with "home-grown" extremist groups developing new capabilities and working together for shared goals. This will be discussed in more detail in chapter 3. New technologies, such as remotely piloted and autonomous air, ground, and sea vehicles; digital fire-control systems for small arms; and machine-learning algorithms that enable image recognition and empower deep fakes and other forms of misinformation, disinformation, and mal-information (MDM), have provided terrorist groups with new capabilities.

2.2 DEFINING TERRORISM

Given the committee's tasking, the appropriate starting point for this assessment is to determine the current state of nuclear terrorism and to assess the risk that a terrorist group, or individual, would conduct a nuclear attack. This is not as straightforward a task as it might first seem, beginning with the challenge of navigating the myriad definitions and typologies of terrorism. (Dolliver and Kearns 2022; Schmid 2004). Even within the U.S. government, agencies use different definitions. After more than two decades after the attacks of 9/11, a consensus definition has proved difficult to achieve. This is partly because some apply the term to express disapproval of a given actor's aims and actions and partly because the lines between terrorism and other forms of violence are often blurred (Wilson 2021).

There are two characteristics, however, that taken together distinguish terrorism from other forms of violence. These characteristics have guided this committee's work. First, terrorists deliberately aim at civilians, which makes terrorism distinct from legal acts of war. Second, terrorists use violence, or the threat of violence, primarily to communicate their support for an ideology, whether religious or political, to generate public attention for their cause. In this way, terrorists are distinct from mass shooters, who generally lack a clear ideological motivation.[1] The target audience for terrorism—the groups the terrorists aim to influence—is larger than the actual victims of the attack, with the goal of achieving a radius of fear that is much greater than the radius of injury and destruction. To accomplish this, terrorists often attack symbols of "enemy" nations or organizations.

2.3 NUCLEAR TERRORISM

The United Nations defines nuclear terrorism as involving the unlawful and intentional use of radioactive material with the intent to cause death, injury, or serious damage to property or the environment, or to compel "a natural or legal person, an international organization or a State to do or refrain from doing an act" (United Nations [UN] 2005). The committee has aligned its assessment of nuclear terrorism to the UN definition by incorporating not just the threat or use of a nuclear explosive device but also the threat or dissemination of nuclear materials (United Nations 2005).

Nuclear terrorism involves the use or threat to use: (1) an illicitly obtained nuclear weapon; (2) an improvised nuclear device (IND); (3) a radiological dispersal device (RDD) commonly referred to as a "dirty bomb" or a radiological exposure device (RED) that exposes unknowing people to radiation; or (4) attacks on nuclear power plants or other facilities that store or use radiological materials. The committee's work covers all of these types along with cyber threats. In addition, the committee looked at how MDM, assisted by artificial intelligence, could both facilitate and amplify the impact of nuclear terrorist attacks. The growing MDM issue is a particular concern for first responders and emergency managers who worry that the public may not follow critical lifesaving guidance during a nuclear incident.

2.4 TOOLS AND TYPES OF NUCLEAR TERRORISM

FINDING 2-1: The possibility that insiders could assist a terrorist in obtaining a state-owned nuclear weapon should not be ruled out. Even if terrorists were unable to overcome use-control safeguards in a stolen weapon, they might still be able to remove fissile material for use in an IND.

2.4.1 State-provided nuclear weapons

A terrorist organization could potentially acquire a stolen nuclear weapon from one of the current nuclear states, either through theft or with insider assistance. Although this is considered highly unlikely (Lieber and Press 2013), it would have the greatest poten-

[1] The "salad bar" phenomenon—which involves violent extremists picking and choosing from a variety of ideologies (e.g., eco-fascism, which endorses the great replacement conspiracy theory embraced by some on the hard right and eco-terrorism adopted by some on the hard left) and "side-switching"—suggests the possibility that terrorists' ideological commitment could be getting somewhat thinner over time (*The Nonprofit Security Grant Program and Protecting Houses of Worship: A View from the American Jewish Community 2022*; The White House 2020. The term "side-switching" was first used by Koehler [2020]).

BOX 2-2
Nuclear and Radiological Nuclear Material

Nuclear and radiological materials are substances that can emit radiation, and these have potential applications in various fields, including energy production, medicine, petrochemical, and military. As materials, they are referred to by the specific isotope from different elements, as different isotopes have difference radiological properties (Friedlander et al. 1981). Understanding the difference between nuclear and radiological materials is crucial in managing and handling these substances effectively (National Academies of Sciences 2021a). This report will use the following definitions for these materials.

Nuclear materials primarily refer to substances that are used in nuclear reactions, such as nuclear power generation, nuclear weapons, or other nuclear technologies (the focus of chapter 6). For this reason, they are often referred to as fissile material. The principal materials of concern that can sustain a nuclear chain reaction, releasing a significant amount of energy, are:

- Uranium-235, a key material for nuclear fission reactions used in nuclear power plants and nuclear weapons.
- Plutonium-239, another crucial material for nuclear fission and often used in nuclear weapons and fuel for nuclear reactors.

Radiological materials emit ionizing radiation, which consists of particles or electromagnetic waves that have sufficient energy to ionize atoms or molecules (see chapter 7). Ionization is the process of removing tightly bound electrons from an atom, resulting in the formation of charged particles (ions). Radiological materials are often used in medical applications, industrial radiography, and certain devices; however, the ionizing radiation can also have dangerous health effects. Different radiological sources include:

- Cobalt-60 and cesium-137 are gamma sources used in medical imaging, cancer treatment, and industrial testing.
- For medicine, technetium-99m is a key material for medical imaging and iodine-131 is a primary treatment for thyroid illnesses and cancer.
- Numerous radioactive tracers are used for various research purposes and well-logging for the petrochemical industry. These include scandium-46, lanthanum-140, manganese-56, sodium-24, antimony-124, iridium-192, iodine-131, silver-110, argon-41, and xenon-133 (International Atomic Energy Agency 2003).

tial for destruction because state-supplied nuclear weapons could be reliable, relatively small, rugged, and transportable, and could have yields up to hundreds of kilotons. One possible scenario is that a nuclear state could make the strategic decision to hand over a nuclear weapon with the aim of having a non-state actor serve as a surrogate. This could potentially include providing the non-state actor with help to override the security and safeguard controls that would otherwise prevent unauthorized or accidental use.

An important deterrent for a state transferring a nuclear weapon to a non-state actor is that if a nuclear weapon is captured or detonated, U.S. nuclear forensics would be able to assess its characteristics and identify or exclude likely contributors (National Academies of Sciences 2021b). This reality could lead a rogue state to try and modify a nuclear weapon or even fabricate a custom device to decrease the likelihood that it would be identified as the source of the device, or to implicate a different nuclear state as the source.

Still another scenario is that a state-owned nuclear weapon could be stolen without assistance from any insiders from a deployment or storage site, or while in transit. In this instance, it is likely that the use controls in the nuclear weapon would prevent the unauthorized use. Although all U.S. nuclear weapons have use controls, little is publicly known about the use and effectiveness of use-control devices in foreign nuclear weapons. If terrorists are unable to overcome the use controls and are thus unsuccessful in detonating the stolen nuclear weapon, they might still be able to extract fissile material for use in an improvised nuclear device. An improvised nuclear device would be designed to generate a nuclear detonation but could have a significantly smaller nuclear yield than state-owned nuclear weapons, although still with catastrophic consequences.

FINDING 2-2: State-on-state warfare, political instability, corruption, or financial crises could provide incentives for a state with nuclear assets to assist a terrorist or terrorist group in fabricating an IND, RDD, or RED.

It is possible that a state might provide terrorists with components, materials, technology, and information needed to fabricate an improvised nuclear or radiological dispersal device. Such assistance would not necessarily have to come from a state that possesses nuclear weapons. Any state that has the ability to produce weapons-usable fissile material and the knowledge to construct a device could provide such assistance. State assistance of any nature would substantially increase the likelihood that a terrorist group could successfully fabricate a reliable improvised nuclear device with some nuclear yield, or a radiological dispersal device.

2.4.2 Improvised Nuclear Device

FINDING 2-3: The most significant barrier to constructing an IND remains the acquisition of a sufficient quantity of fissile material. Even if this barrier were overcome, an IND would be less reliable, produce a less predictable yield, and be more vulnerable to accidental detonation than state-stockpiled weapons. Nevertheless, the political and psychological impact of any nuclear detonation would be consequential.

Improvised Nuclear Devices are devices that are constructed using diverted or stolen weapons-usable fissile material. Concern about the IND risk has been raised publicly as far back as 1977 when the U.S. Congress's Office of Technology Assessment noted that: "A small group of people could possibly design a crude nuclear explosive device (assuming) sufficient quantities of fissile material have been provided" (Office of Technology Assessment 1977). There is a remote potential that fissile material could be harvested from a stolen nuclear weapon, but the more probable scenario is that it could be provided with insider assistance. There is no lack of open-source information on improvised nuclear devices, although the accuracy of the origination can be suspect. That said, possessing the expertise is not the most significant barrier to constructing an IND. Instead, it is the challenge of obtaining enough fissile material to sustain a nuclear chain reaction. This underscores the importance of global efforts to reduce the stocks of HEU, separated plutonium, and other fissile materials and bolstering the security of the materials that remain. The DOE/NNSA had a long-running Materials Protection, Control, and Accounting (MPC&A) Program designed to improve the localization, tracking, and control of those materials (National Nuclear Security Administration 2001; U.S. Government Accountability Office 2020).

Given the many challenges involved, an IND is likely to be unreliable (Ferguson et al. 2005; Langewiesche 2006), have a low and unpredictable yield compared with sophisticated high-yield nuclear weapons developed by governments, and be vulnerable to accidental detonation. Nevertheless, the consequences of detonating an IND in a major city would still be catastrophic, potentially killing tens or hundreds of thousands of people and wounding hundreds of thousands more.

2.4.3 Radiological Dispersal Devices and Radiological Exposure Devices

FINDING 2-4: The technical barriers to constructing and employing an RDD or RED are much lower than they are for a nuclear weapon or IND, making for a greater likelihood that terrorists will gain access to the means to disperse radiological materials.

Two types of potential devices that terrorists might use that would not produce a nuclear yield but still have significant effects are the RDD and the RED. The successful employment of an RDD or an RED would have significantly less destructive potential than a nuclear explosion but would still be effective in causing significant economic, physical, and psychological damage and could include loss of life. The source materials for an RDD or an RED are more available and often less secure than the material needed for an IND, and thus the probability of use is higher. In addition, the technical knowledge needed to fabricate these devices is much lower than that needed for an IND.

The RDD is often referred to as a "dirty bomb." It uses propellants, explosives, or other means to disperse radiological materials. An RDD could be constructed as a complete device, containing explosive/propellant mated to radiological material, or implemented as an explosive device detonated in close proximity to a source of radiological material, either stationary or in transit.

An example of a dirty bomb scenario took place in 1995, when Chechen leader Shamil Basayev threatened to detonate radioactive containers in Russian cities. During an interview in November 1995, Basayev told a Russian television news crew where such a device was buried in a Moscow park. Russian authorities subsequently found a small explosive package with cesium-137 buried a few inches below the ground. Basayev told the reporters that he wanted to demonstrate that Chechen rebels could conduct such an attack and that he intentionally disclosed the location as a "small disarmament" signal (Bale 2004; Specter 1995).

A detonated RDD would impact local health, inflict significant economic damage, and potentially trigger widespread panic. The impacted area could be several city blocks and would be closed off to the public during the resulting extensive and costly cleanup. Some loss of life cannot be ruled out, but it would vary depending on the proximity of people to the explosion as well as its size and the radiological materials employed. The economic effect could be prolonged, well after the cleanup was completed, as a result of lingering public anxiety and fear of residual radiation exposure. If explosives or propellants are employed, it is likely that the RDD would be identified as such relatively quickly, allowing time to mitigate any potential health issues.

An RED exploits existing radiological material or objects containing radiological material. It is intended to expose people to radiation without their knowledge and with the intent of generating significant health effects for those exposed. An RED would not be as obvious as explosive RDD and could remain in place for an extended period of time. Detecting an RED would be possible by a positive reading on radiation detection equipment; or, if the object was in place for an extended period of time, it might be discovered only after individuals seek medical help for radiation-induced health issues such as skin reddening or acute radiation sickness. Detection via the latter method would be very slow and challenging.

Reducing the amount of radiological material and putting in place strong safeguards can contribute to reducing the RDD and RED risk. However, despite NNSA's ongoing programs, such as the Off-Site Source Recovery Project (OSRP) (Coel-Roback 2019), commercial and medical radiological materials remain widely distributed throughout the world. Some of these have been abandoned, and others are under little or no control. RDDs also have often been described as "weapons of mass disruption" because their impact would be primarily psychological and economic (U.S. Department of Health and Human Services 2023). Deployment of an RDD at an economic choke point, such as a major seaport or transportation hub, could have outsized economic and social impact. The impact of an RED would be very different given the potential for long periods of public exposure to radiation in the impacted area along with the possibility of adverse health effects. Nevertheless, long-term economic damage is possible as a result of a lingering fear of radiation exposure even after the object was removed and residual radioactivity cleaned up.

2.4.4 Attack or Sabotage of Nuclear Facilities

FINDING 2-5: Based on intelligence information in government reports and open-source literature, there have been instances of sabotage at nuclear facilities by knowledgeable insiders, but to date they have been rare.

Vulnerabilities at nuclear facilities can make them susceptible to malicious acts and create opportunities for terrorists or criminal groups. Nuclear facilities are protected through redundant layers of physical protection measures designed to prevent access to the nuclear facility or nuclear material. Additional tools such as remote monitoring, security guards and response forces, and intelligence are also used. (International Atomic Energy Agency 2023b). Nuclear facilities that require physical protection include nuclear reactors, fuel cycle facilities, and spent fuel storage and disposal facilities. Although site- and event-specific threat assessments are used to determine how much physical protection is required, common elements include physical protection areas that are graded to provide defense-in-depth with barriers and controls for the exclusion area, protected area, vital area, and material access areas. Physical protection systems include intrusion detection systems to notify the site security force of a potential intruder, typically with intrusion alarm assessment systems to help distinguish false or nuisance alarms from actual intrusions. An armed response capability may be necessary to defend nuclear material or a nuclear facility against an intrusion or attack and to protect public health and safety, depending on the nature and quantity of the material. In addition, local, state, and federal agencies may be called on to provide off-site assistance in an emergency at a U.S. facility (Nuclear Regulatory Commission 2020).

Sabotage at nuclear facilities by knowledgeable insiders has occurred, but reports of such incidents have been rare. The ultimate aim of sabotage would be to damage the facility, causing a radiological release harmful to nearby populations. Nuclear facilities use multiple physical and other systems to prevent sabotage, with safety-critical systems subject to the highest levels of engineered and administrative controls by the facility operators. Examples of non-physical systems include employee behavioral observation programs, psychological testing, fitness for duty criteria, and measures such as "two-person or three-person rules" requiring that two or three operators must be present and responsible for certain high-consequence activities (World Nuclear Association 2022).

A recent NAS report has identified that "transportation is the most vulnerable phase in protecting nuclear materials from a security standpoint, as the material is removed from the confines of the nuclear facility" (National Academies of Sciences 2023). This is discussed in more detail in chapter 8. Regulatory, technical, and operational considerations are part of an overall strategy for providing secure transport of fresh and spent nuclear fuel. This is particularly important when transporting materials across international boundaries, as countries have different regulations based on their level of risk acceptance. Internationally, the IAEA provides umbrella guidance for the Convention on the Physical Protection of Nuclear Materials (CPPNM) obligations for materials in transport (International Atomic Energy Agency 2023a). Spent fuel security requirements for U.S. domestic transportation are defined in 10 CFR73.37, which provides detailed requirements that include development of a security plan with the full range of protective components needed to properly protect shipments during transport (National Academies of Sciences 2023).

2.4.5 Cyberattacks on Nuclear Facilities

FINDING 2-6: Intelligence analysts and journalists are reporting that there is a growing number of attempted cyberattacks on critical infrastructure targets, including nuclear facilities, varying in degree of sophistication and threat, and originating from both state and non-state actors.

As with other industrial and infrastructure facilities, nuclear facilities rely on information and communication systems, thus making them potentially vulnerable to cyberattacks. At U.S. commercial nuclear facilities, information and communication systems are routinely evaluated for cyber risks and vulnerabilities. To reduce further the possibility of a successful cyberattack, the utilization of such systems within vital areas of the facility is strictly regulated by the NRC and state regulatory agencies.

As with the case of all infrastructure targets and the economic sector, cyberattacks on nuclear facilities are increasing in number and sophistication. State and non-state actors routinely probe the security of critical infrastructure systems to find vulnerabilities that will permit access. In the relatively small number of publicly reported attacks where adversaries have succeeded in penetrating cyber defenses for nuclear facilities, postmortems have documented the consequences to date to be limited to exfiltrated employee information, corporate data, and non-sensitive technical data.

Operators of U.S. commercial nuclear facilities must notify the U.S. Nuclear Regulatory Commission within one hour of discovering a cyberattack that adversely impacted safety, security, support systems, or emergency preparedness functions (including offsite communications). Longer notification timelines are allowed for successful attacks of lesser severity (Nuclear Regulatory Commission 1973).

Ransomware attacks have also occurred in a small number of cases worldwide. Some of these attacks have succeeded in penetrating and temporarily disabling plant monitoring systems, although with no deleterious effects on plant operations or resultant radiological release (World Nuclear Association 2022).

2.4.6 Emerging Threat: Misinformation, Disinformation, and Mal-information

FINDING 2-7: Nuclear terrorists can tap into the widespread dread of radiation and nuclear weapons to generate significant consequences by taking advantage of MDM. New developments in AI could make MDM even more damaging.

Both the 2021 National Strategy for Countering Domestic Terrorism and the 2022 National Security Strategy highlight information operations as threats to U.S. national security (The White House 2020, 2021). Terrorists have long used information operations to recruit supporters and frighten their enemies. Information operations could be especially powerful in the context of nuclear terrorism, in part because people generally fear radiation and nuclear weapons (Stern 1999). The members of this committee are particularly concerned about the ways in which artificial intelligence could be used to amplify underlying public fears associated with nuclear risk.

While MDM has been used throughout history, the creation of new digital platforms has expanded the global means of communication and connections, thereby allowing users to exchange information quickly and widely (McBride et al. 2021). False information spreads more quickly than the truth (Vosoughi, Roy, and Aral 2018), and countering falsehoods is complicated by the fact that a majority of American adults are now getting their news from digital platforms. While some platforms are responsible, others have few to no safeguards to prevent use by nefarious actors including terrorists (Shearer 2021). This provides ample opportunities to expose large swaths of the American public to harmful MDM.

MDM is a complex and nuanced phenomenon that can take many forms, including news media and social media posts. Nation-state adversaries are known to use MDM to amplify extremist ideologies, using MDM campaigns to cast doubt on official narratives, amplify political discord, spark confusion, and promote favorable narratives surrounding themselves, their allies, nonaligned countries, or certain domestic actors. For example, China, Russia, and Iran systemically amplified Q-Anon ideology on social media leading up to the January 2021 Capitol insurrection. One-fifth of all Q-Anon posts on Facebook in 2020 originated overseas (The Soufan Center 2021). These campaigns are part of broader efforts by adversaries to negatively impact the credibility and functioning of government (Colomina, Sánchez Margalef, and Youngs 2021). Their goal is to exacerbate societal divisions and grievances, spark civil unrest and violence, and degrade the operational security and strategic functioning of the targeted country's defense structure (Wolters et al. 2021). Given the open and widespread nature of the media environment, MDM has the potential to be used by a wide variety of terrorists, including nuclear terrorists.

The terms misinformation, disinformation, and mal-information are often used interchangeably, despite each term having its own distinct meaning. According to Wolters, the difference between the terms misinformation and disinformation lies in the intent of the content's creator (Wolters et al. 2021). *Misinformation* occurs when the creator has no intention to deceive, yet still shares false information thinking it's true (Wolters et al. 2021). *Disinformation* is when the creator shares false information with the goal of deceiving the consumer. For a piece of information to be considered disinformation, the user must know the information is false and be looking to intentionally mislead, harm, or manipulate (Wolters et al. 2021). *Mal-information*, however, is different from misinformation and disinformation in that it is based in fact (but can contain some false elements as well) yet is spread with the intent to cause harm. An example of mal-information can include leaks of personal or private information meant to damage the reputation or compromise the safety of an individual (Wolters et al. 2021).

While all three concepts are critical to understanding how the information space can be weaponized, intentionally spreading false information in the context of nuclear terrorism could significantly impede an emergency response. Even poorly designed disinformation campaigns could impact confidence in government institutions and reputable journalistic outlets (Wolters et al. 2021), making it an extremely useful tool for nuclear terrorists.

With the ability of MDM to amplify public anxieties, a nuclear or radiological device could have minimal actual destructive effect and yet have far-reaching psychological and political impact (Ackerman 2008). Studies demonstrate that humans pay more attention to information that arouses emotion—especially rage, fear, surprise, and disgust—making MDM's use in the context of nuclear terrorism potentially devastating (Wolters et al. 2021). Indeed, AI-enhanced nuclear threats could be enough to spark widespread public fear and panic (Giorgidze and Wither 2019; Johnson 2022) analogous to the widespread fear generated by the 1938 *War of the Worlds* radio broadcast.

Terrorists could use targeted disinformation campaigns to encourage individuals working in secure locations, including at nuclear-related facilities, to engage in acts of sabotage. A potentially sobering scenario is one where a terrorist group uses MDM to try and provoke a nuclear weapons state to use a nuclear weapon. For example, in 2016, Pakistani defense minister Khawaja Muhammad Asif threatened to use nuclear weapons against Israel after reading a fake news article claiming that Israel planned to "destroy [Pakistan] with a nuclear attack" if Pakistan sent troops to Syria (Goldman 2016). While this situation was peacefully resolved, the risk may grow as the methods and sophistication of MDM evolve.

Another established motivation for terrorists is to generate attention. Former British prime minister Margaret Thatcher once observed that "publicity is the 'oxygen' of terrorism" (Thatcher 1985). As Gary Ackerman (2008) has pointed out, a nuclear attack would undoubtedly provide unrivaled attention to a terrorist and their cause.

If a nuclear attack or accident were to occur, MDM could jeopardize the credibility of risk mitigation and response teams and discourage the public from cooperating or seeking help. This could also apply to public health responses following a nuclear disaster that are undermined by MDM akin to what impaired COVID-19 vaccination efforts (United Nations Interregional Crime and Justice Research Institute [UNICRI] 2022). Lastly, continuous victimization could occur through media campaigns that show pictures or videos of the devastation, which can be doctored to augment and prolong a society's trauma.

While the practical impact of MDM on nuclear terrorism remains to be seen, its use by both nation-state adversaries and non-state actors poses a significant threat to American and global security. It has led to the creation of groups working to combat MDM such as the DHS Cybersecurity and Infrastructure Security Agency's MDM team, the FBI's Foreign Influence Task Force, and the ODNI's Foreign Malign Influence Center (U.S. Cyberspace Solarium Commission 2021).

The distrust of government institutions by a growing segment of the U.S. population could manifest in a disinformation campaign designed to undermine government and scientific credibility and instill chaos and panic during a nuclear incident. Ideally there would be a cadre of trusted nongovernmental scientific and technical experts at the state and local levels to augment national experts. If these nongovernment experts were trained in public communications, they would be in a position to assist governors, mayors, and tribal and territorial leaders in getting lifesaving information out during major emergencies. Polls generally show high levels of public trust in communications from health, scientific, technical, and educational professionals that are locally based. (See, for example, Ferriman 2001; Kennedy, Tyson, and Funk 2022; and Yi and Sawyer 2021). Exercises sponsored by DHS could involve these experts, and they could also be incorporated into state and local emergency management plans.[2]

RECOMMENDATION 2-1: The Department of Homeland Security—with support from the Centers for Disease Control and Prevention, the National Governors Association, and the U.S. Conference of Mayors—should undertake a multipronged effort involving all levels of government (federal, state, local, tribal, and territorial) to include research and educational entities, civic associations, and media to raise public awareness and understanding of how information can be used to confuse, mislead, and deceive during major crises.

The effort should actively engage civil society and could be modeled on a recent program in Finland that strengthens the ability of the public to critically analyze the information they receive from all sources (U.S. Cyberspace Solarium Commission 2021). As discussed in chapter 9, the United States at state and local levels needs to capitalize on the many sources of experts that can be marshaled to create a whole-of-country capability.

References

Ackerman, Gary. 2008. "Motivations for Engaging in Nuclear Terrorism." *FfP Threat Convergence Expert Series*. https://www.files.ethz.ch/isn/46247/Motivations%20for%20Engaging%20in%20Nuclear%20Terrorism.pdf.

Albright, David. 2010. *Peddling Peril: How the Secret Nuclear Trade Arms America's Enemies*. Free Press.

Bale, Jeffrey. 2004. "The Chechen Resistance and Radiological Terrorism." Nuclear Threat Initiative. https://www.nti.org/analysis/articles/chechen-resistance-radiological-terror/.

Coel-Roback, Rebecca J. 2019. *Off-Site Source Recovery Program* (United States). https://www.osti.gov/biblio/1565825.

Colomina, Carme, Héctor Sánchez Margalef, and Richard Youngs. 2021. *The Impact of Disinformation on Democratic Processes and Human Rights in the World*. European Parliament. https://www.europarl.europa.eu/RegData/etudes/STUD/2021/653635/EXPO_STU(2021)653635_EN.pdf.

Dolliver, Matthew J., and Erin M. Kearns. 2022. "Is It Terrorism? Public Perceptions, Media, and Labeling the Las Vegas Shooting." *Studies in Conflict & Terrorism* 45 (1): 1–19. https://doi.org/10.1080/1057610X.2019.1647673.

[2] After the September 11, 2001, attacks, the Department of Homeland Security funded the creation of an "Infrastructure Experts Team," managed by Oak Ridge Associated Universities. The team, composed of about two dozen faculty from universities in the Washington, DC, area, conducted table-top exercises and carried pagers to respond to a terrorist attack or other emergency on short notice.

Federal Emergency Management Agency. 2022. *Planning Guidance for Response to a Nuclear Detonation Third Edition*. Edited by Chemical, Biological, Radiological, and Nuclear (CBRN) Office

Ferguson, Charles D., William C. Potter, Amy Sands, Leonard S. Spector, and Fred L. Wheling. 2005. *The Four Faces of Nuclear Terrorism*. Routledge.

Ferriman, Annabel. 2001. "Poll Shows Public Still Has Trust in Doctors." *BMJ* 322 (7288): 694. https://doi.org/10.1136/bmj.322.7288.694. https://www.ncbi.nlm.nih.gov/pmc/articles/PMC1173253/.

Friedlander, Gerhart, Joseph W. Kennedy, Edward S. Macias, and Julian M. Miller. 1981. *Nuclear and Radiochemistry*. Wiley.

Giorgidze, Lasha, and James K. Wither. 2019. "Horror or Hype: The Challenge of Chemical, Biological, Radiological, and Nuclear Terrorism." *Occasional Paper Series*. The George C. Marshall European Center for Security Studies, December 2019, Number 031. https://www.marshallcenter.org/en/publications/occasional-papers/horror-or-hype-challenge-chemical-biological-radiological-and-nuclear-terrorism-0

Goldman, R. 2016. "Reading Fake News, Pakistani Minister Directs Nuclear Threat at Israel." *New York Times*, December 24, 2016. https://www.nytimes.com/2016/12/24/world/asia/pakistan-israel-khawaja-asif-fake-news-nuclear.html.

Graham, B., J. Talent, G. Allison, R. Cleavaland, S. Rademaker, T. Roemer, W. Sherman, H. Sokolski, and R. Verma. 2008. *World at Risk: The Report of the Commision on the Prevention of WMD Proliferation and Terrorism*. Random House.

International Atomic Energy Agency. 2003. *Radiation Protection and the Management of Radioactive Waste in the Oil and Gas Industry*. Safety Reports Series. International Atomic Energy Agency.

---. 2023a. "Convention on the Physical Protection of Nuclear Material (CPPNM) and Its Amendment." https://www.iaea.org/services/key-programmes/smr-platforms-nhsi.

---. 2023b. "Security of Nuclear and Other Radioactive Material." https://www.iaea.org/topics/security-of-nuclear-and-other-radioactive-material.

Johnson, James. 2022. "Inadvertent Escalation in the Age of Intelligence Machines: A New Model for Nuclear Risk in the Digital Age." *European Journal of International Security* 7 (3): 337–359. https://doi.org/10.1017/eis.2021.23.

Kennedy, Brian, Alec Tyson, and Cary Funk. 2022. "Americans' Trust in Scientists, Other Groups Declines." *Trust in Science*. https://www.pewresearch.org/science/2022/02/15/americans-trust-in-scientists-other-groups-declines/.

Koehler, Daniel. 2020. "Switching Sides: Exploring Violent Extremist Intergroup Migration across Hostile Ideologies." *Political Psychology* 41 (3): 499–515. https://doi.org/https://doi.org/10.1111/pops.12633.

Langewiesche, William. 2006. "How to Get a Nuclear Bomb." December 2006. https://www.theatlantic.com/magazine/archive/2006/12/how-to-get-a-nuclear-bomb/305402/.

Lieber, Keir A., and Daryl Press. 2013. "Why States Won't Give Nuclear Weapons to Terrorists." *Quarterly Journal: International Security* 38 (1): 80–104.

McBride, M., H. Wolters, K. Haney, and W. Rosenau. 2021. *The Psychology of (Dis)information: Case Studies and Implications*. CNA. https://www.cna.org/reports/2021/10/The-Psychology-of-%28Dis%29information-Case-Studies-and-Implications.pdf.

Mowatt-Larssen, Rolf. "Al Qaeda Weapons of Mass Destruction Threat: Hype or Reality?." Belfer Center for Science and International Affairs, Harvard Kennedy School, January 2010. https://www.belfercenter.org/publication/al-qaeda-weapons-mass-destruction-threat-hype-or-reality

---. 2020. *A State of Mind: Faith and the CIA*. BookBaby.
National Academies of Sciences, Engineering, Medicine. 2021a. *Radioactive Sources: Applications and Alternative Technologies*. Washington, DC: The National Academies Press.
---. 2021b. *Restoring and Improving Nuclear Forensics to Support Attribution and Deterrence: Public Summary*. Washington, DC: The National Academies Press.
---. 2023. *Merits and Viability of Different Nuclear Fuel Cycles and Technology Options and the Waste Aspects of Advanced Nuclear Reactors*. Washington, DC: The National Academies Press.
National Nuclear Security Administration. 2001. "MPC&A Program Strategic Plan." https://media.nti.org/pdfs/32-nnsa.pdf.
National Research Council. 1996. *An Assessment of the International Science and Technology Center: Redirecting Expertise in Weapons of Mass Destruction in the Former Soviet Union*. Washington, DC: The National Academies Press.
The Nonprofit Security Grant Program and Protecting Houses of Worship: A View from the American Jewish Community. 117th (2022). https://www.congress.gov/117/chrg/CHRG-117jhrg47469/CHRG-117jhrg47469.pdf
Nuclear Regulatory Commission. 1973. *10 CFR Part 73—Physical Protection of Plants and Materials*.
---. 2020. "Physical Protection." Last modified March 11, 2020. https://www.nrc.gov/security/domestic/phys-protect.html.
Office of Technology Assessment. 1977. *Nuclear Proliferation and Safeguards*.
Rotblatt, Joseph. 1998. *Nuclear Weapons: The Road to Zero*. Boulder, CO: Westview Press.
Schmid, Alex. 2004. "Terrorism—The Definitional Problem." *Case Western Reserve Journal of International Law* 36 (2). https://scholarlycommons.law.case.edu/jil/vol36/iss2/8.
Shearer, Elisa. 2021. "More than Eight-in-Ten Americans Get News from Digital Devices." https://www.pewresearch.org/short-reads/2021/01/12/more-than-eight-in-ten-americans-get-news-from-digital-devices/.
Specter, Michael. 1995. "Chechen Insurgents Take Their Struggle to a Moscow Park." *New York Times*, November 24, 1995.
Stern, Jessica. 1999. *The Ultimate Terrorists*. Harvard University Press.
Thatcher, Margaret. 1985. "Speech to American Bar Association ('We Must Try to Find Ways to Starve the Terrorist and the Hijacker of the Oxygen of Publicity on Which They Depend')." https://www.margaretthatcher.org/document/106096.
The Soufan Center. April 2021. *Quantifying the Q Conspiracy: A Data-Driven Approach to Understanding the Threat Posed by QAnon*. The Soufan Center. https://thesoufancenter.org/wp-content/uploads/2021/04/TSC-White-Paper_QAnon_16April2021-final-1.pdf.
The White House. 2021. *National Strategy for Countering Domestic Terrorism*.
---. 2020. National Security Strategy. *Threats to the Homeland*. 116th (2020) https://www.govinfo.gov/content/pkg/CHRG-116shrg42870/html/CHRG-116shrg42870.htm
U.S. Congress. 2007. *Implementing Recommendations of the 9/11 Commission Act of 2007*.
U.S. Cyberspace Solarium Commission. 2021. *Countering Disinformation in the United States - CSC White Paper #6*. https://cybersolarium.org/wp-content/uploads/2022/05/COUNTE1.pdf.
U.S. Department of Health and Human Services. 2023. "Radiation Emergency Medical Management: Radiological Dispersal Devices (RDDs)." Last modified January 19, 2023. https://remm.hhs.gov/rdd.htm.
U.S. Government Accountability Office. 2020. Nuclear Nonproliferation: Past U.S. Involvement Improved Russian Nuclear Material Security, but Little Is Known about Current Conditions.
United Nations. 2005. International Convention for the Suppression of Acts of Nuclear Terrorism.

United Nations (UN). 2005. *International Convention for the Suppression of Acts of Nuclear Terrorism*. https://treaties.un.org/doc/db/terrorism/english-18-15.pdf.

United Nations Interregional Crime and Justice Research Institute (UNICRI). 2022. *Handbook to Combat CBRN Disinformation*. Torino, Italy.

Vosoughi, Soroush, Deb Roy, and Sinan Aral. 2018. "The Spread of True and False News Online." *Science* 359 (6380): 1146–1151. https://doi.org/10.1126/science.aap9559. https://www.science.org/doi/abs/10.1126/science.aap9559.

Wilson, Chris. 2021. "The Ubiquity of Terror: Politically Motivated Killing of Civilians." *Nationalism and Ethnic Politics* 27 (3): 389–404. https://doi.org/10.1080/13537113.2021.1920709.

Wolters, H., K. Stricklin, N. Carey, and M. McBride. 2021. *The Psychology of (Dis)information: A Primer on Key Psychological Mechanisms*. Center for Naval Analyses. https://www.cna.org/reports/2021/10/The%20Psychology-of-%28Dis%29information-A-Primer-on-Key-Psychological-Mechanisms.pdf.

World Nuclear Association. 2022. "Security of Nuclear Facilities and Material." Last modified March 2022. https://www.world-nuclear.org/information-library/safety-and-security/security/security-of-nuclear-facilities-and-material.aspx.

Yi, Jinhee, and Johnny Sawyer. 2021. "Americans Are Most Likely to Trust Healthcare Workers and Doctors." *Game Changers*. https://www.ipsos.com/en-us/americans-are-most-likely-trust-healthcare-workers-and-doctors.

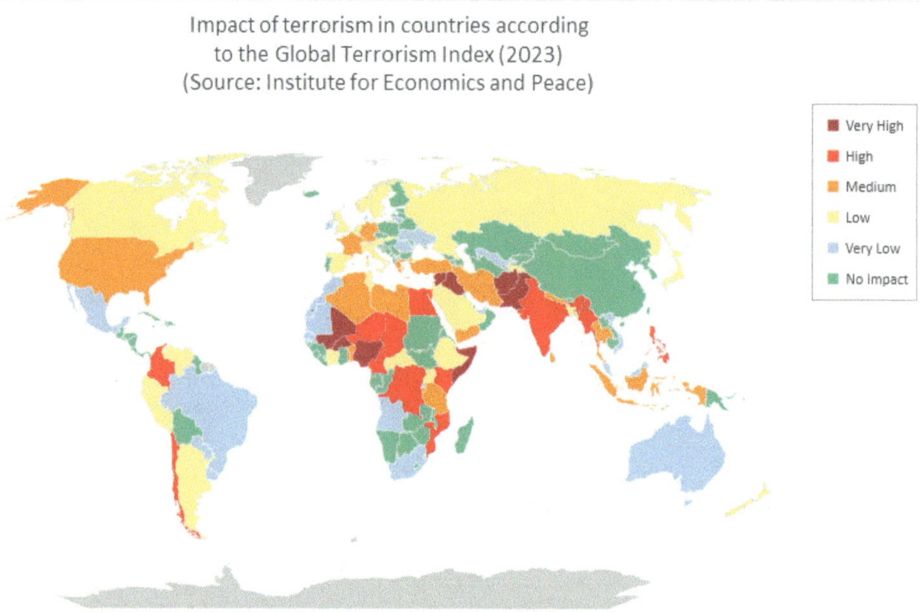

FIGURE 3-1 The Global Terrorism Index (GTI) is a comprehensive study analyzing the impact of terrorism for 163 countries, produced by the Institute for Economics & Peace (IEP) using data from Terrorism Tracker and other sources. Shown here are countries and levels of impact terrorism has, ranging from very high to no impact. The map is presented as a reminder of the continuing activities of international and domestic terrorists and, as this chapter discusses, how the distinction between domestic and international terrorism is becoming blurred.
NOTE: The GTI defines terrorism as "the systematic threat or use of violence, by non-state actors, whether for or in opposition to established authority, with the intention of communicating a political, religious or ideological message to a group larger than the victim group, by generating fear and so altering (or attempting to alter) the behaviour of the larger group." The GTI recognizes that many states commit terror against their citizens. Acting under the authority as a state is not included in the GTI.
SOURCE: Institute for Economics and Peace 2023.

3

New Dynamics in Nuclear Terrorism Pose New Risks

3.1 TERRORISTS' LONG-STANDING INTEREST IN NUCLEAR AND RADIOLOGICAL WEAPONS

Terrorist groups have been pursuing nuclear and radiological weapons and devices for decades (Ferguson et al. 2005). A few prominent examples include: (1) Aum Shinrikyo's unsuccessful, repeated attempts to mine uranium, hire Russian nuclear scientists, and purchase Russian nuclear technology in the 1990s; (2) Al Qaeda's unsuccessful alleged attempts to purchase nuclear weapons and material from a variety of sources in the 1990s and recruit insiders, as well as attempts to design a nuclear weapon that would have included explosive testing in Afghanistan; and (3) ISIS's access to a cache of cobalt-60 when it overran Mosul (Ward 2018). In August 2014, an insider at the Doel 4 nuclear power plant in Belgium carried out a successful act of nuclear sabotage. The sabotage did not result in the release of radiological material, but the costs were significant. The saboteur was never identified, but the nuclear regulator added new requirements for defending against insider threats (Bunn et al. 2016).

Although terrorists confront considerable challenges in the pursuit of a nuclear weapon or an improvised nuclear device (IND), the enormous consequences of a nuclear detonation call for continued vigilance by the United States and reinvigorated efforts by the international community to reduce the risk. Attacks involving nuclear sabotage or deploying a radiological dispersal device (RDD) have fewer barriers for non-state actors to pursue. These risks reinforce the need for an ongoing international commitment to managing and reducing nuclear terrorism challenges.

Radiological material found in secure nuclear power plants and other facilities can be used to make an RDD (aka "dirty bomb") or a radiological exposure device (RED) (Council on Foreign Relations 2006). According to the National Consortium for the Study of Terrorism and Responses to Terrorism (START) at the University of Maryland, there have been over 80 attacks on nuclear facilities around the globe since 1963, the

BOX 3-1
Summary

The probability of nuclear terrorism is low, due in part to a number of programs, policies, and systems developed to secure and control access to nuclear weapons and weapons-usable materials. Nuclear terrorism is also hindered by the challenges non-state actors must overcome to successfully fabricate an improvised nuclear device (IND). Nonetheless, the number and types of groups who are motivated to try to use an IND or radiological dispersal device/ radiological exposure device is likely growing. While some non-state actors are more likely to be deterred by the near-certainty of attribution and retribution, others, including millenarian groups such as ISIS and U.S.-based accelerationist groups, actively court retaliation to spark a wider war or to realize apocalyptic beliefs (Earnhardt, Hyatt, and Roth 2021).

The risk of nuclear terrorism must also be evaluated in the context of changing norms and a potential renaissance in civil nuclear power. There are also the mounting dangers associated with cyberattacks on operating nuclear power plants. Russia has demonstrated a willingness to defy international norms by attacking and occupying Ukraine's operating civilian nuclear power plants. Russia has also been employing proxies with a history of war crimes and has deployed operatives to attack and poison individuals with advanced nerve agents and radiological substances.

A particularly troubling development is the increase in the domestic terrorism risk within the United States, which may also increase the risk of nuclear terrorism. This includes U.S.-based accelerationist groups who have been deliberately recruiting U.S. military personnel and have targeted critical infrastructure including nuclear facilities. Additionally, there are disturbing and growing U.S. domestic links

majority of which are understood to have been orchestrated by malicious actors (as distinct from protestors) (Ackerman 2023). In one notable instance, Belgian police discovered that individuals connected to ISIS secretly monitored a senior nuclear industry official who worked at nuclear access sites that contained enough highly enriched uranium for several nuclear bombs (Bunn 2016).

3.2 NEW ERA IN NUCLEAR TERRORISM

FINDING 3-1: A new nuclear era is emerging, in which (1) long-standing arms control treaties as well as counterproliferation and nonproliferation norms and arrangements are eroding, (2) civil nuclear energy is becoming more widespread and utilizing new designs, and (3) the strategic focus on combating international terrorist activity is waning. As evidenced in the 2022 invasion of Ukraine by the Russian military, attacks on Ukrainian nuclear power plants signal the end of a long-standing norm that such plants should be inviolate.

with mercenary and terrorist groups across international borders. Extremists are utilizing social media to fuel radicalization and political polarization, as well as to propagate dis- and misinformation, and sow mistrust of government institutions and authoritative information. Social media is serving as a powerful organizational tool for terrorist groups, facilitating an increase in international connectivity among domestic and foreign terrorist organizations.

In sum, managing the threat of nuclear terrorism will be challenged by the continued presence of international and domestic terrorist groups, some of whom are motivated to carry out these kinds of attacks.

Highlights

- A new era of potential nuclear instability is emerging with the demise of traditional arms control agreements, and challenges to some long-standing nonproliferation arrangements and norms.
- Russian attacks on and occupation of Ukrainian power plants signal the end of a well-established norm that such plants should be inviolate.
- The demarcation between domestic and international terrorist organizations is blurring alongside the use of non-state actors as proxies by states. Terrorism is increasingly transnational. This has strategic, organizational, and operational implications for U.S. security efforts that assign roles and responsibilities based on whether a threat is state versus non-state and domestic versus international.
- Extreme-right-wing accelerationist groups are recruiting U.S. military personnel, potentially posing an increased risk to security across the nuclear weapons enterprise and to the civil nuclear industry.

Recent Russian attacks on Ukrainian nuclear power plants are eroding the well-established norm that such plants should be inviolate. Ukrainian president Volodymyr Zelensky has accused Russian president Vladimir Putin of "nuclear terrorism" for his attack and seizure of the Zaporizhzhia nuclear power plant (Broad 2022). Zelensky also called out Russia for conducting missile strikes that hit the industrial equipment surrounding the Pivedennoukrainsk nuclear site (Ritter and Gambrell 2022). In occupying the Chernobyl nuclear site, Russian forces reportedly damaged power supplies that were necessary to prevent releases of radiation (Kuleba 2022).

Russia's attacks on nuclear sites have also occurred in cyberspace, where international norms against attack are less established. In March 2022, the U.S. Department of Justice unsealed charges on four Russian officials accused of hacking into the computer system of Wolf Creek Nuclear Operating Corporation in Kansas (Benner and Conger 2022). No sabotage of the facility resulted from these cyberattacks, but it is troubling that Russia would undertake an attack, possibly intended to disrupt or compromise operations, on a nuclear facility (Nuclear News 2022; Swan and Miller 2022).

There is also the ongoing risk that nuclear states may lose control of the weapons in their possession. This is not only a risk for states such as Pakistan and North Korea. During the Cold War and prior to 1968, the United States lost six nuclear weapons or components that have never been recovered (Roza 2022; Suciu 2021), and the Soviet Union may have lost more (Federation of American Scientists 2023).

Acts of nuclear terrorism ultimately require fulfillment of three elements: (1) procurement of or access to nuclear weapons or radiological materials; (2) capability to steal, or develop, and deploy a nuclear weapon, IND, or RDD/RED; and (3) the motivation to undertake such an attack. Each of these is discussed below:

- Access: To be successful, terrorists must procure some or all of the following:
  - nuclear weapon(s) from a state actor or an insider; or
  - financial or technical resources sufficient to acquire nuclear weapons, components, or fissile/radiological materials; or
  - insider access to, or access to third-party insiders capable of supplying nuclear and/or radiological materials (plutonium, highly enriched uranium (HEU), high-activity radiological sources); and
  - a sanctuary, technical facility, and sufficient time to design, develop, and prepare the weapon.
- Capabilities:
  - access to appropriately trained technical experts (such as metallurgists, engineers, chemists, and physicists);
  - operational security (i.e., managing operations such that countermeasures by a state or the international community are less likely to identify and interrupt the activity);
  - knowledge of or access to insiders capable of supplying the knowledge base for producing, handling, and potentially testing an IND, RDD, or RED;
  - covert transportation of material and equipment and other logistical support; and
  - means of targeting and delivery.[1]
- Possible motivations include:
  - incite civil war;
  - attract media attention to the cause;
  - create fear and chaos;
  - generate mass economic disruption and destructive cascading effects—for example, by targeting critical systems, such as ports, power and water infrastructure, or supply systems;

[1] Transporting a nuclear device over long distances to a targeted location in the United States is problematic. Terrorists would likely prefer to maintain complete control over the device, which means they would likely avoid major ports or border crossings to smuggle the device. However, if an adversary's intent is to cause mass disruption to the intermodal transportation system, they may decide to send a nuclear device in a cargo container with the goal of targeting the arriving port facility.

- religious goals such as to influence deities or timing of apocalypse;
- promote opposition to nuclear power by attacking nuclear energy facilities; and
- demonstrate technical prowess to sympathizers as well as the targeted group.

3.3 DOMESTIC AND INTERNATIONAL TERRORISM: BLURRING BOUNDARIES

FINDING 3-2: The trends of the past years have demonstrated that domestic and international terrorist organizations are becoming more closely linked and difficult to differentiate. Countering these increasingly transnational organizations through close interagency and international cooperation will be challenged by the barriers associated with differing jurisdiction, authorities, and capacity along with the varying missions of the intelligence and law enforcement communities.

Importantly, the risk of nuclear terrorism is probably greatest where the lines between domestic and international terrorists and between state and non-state actors is most unclear. The U.S. government divides terrorists into two categories: domestic and international. According to the FBI's definitions, "international terrorists" are individuals or groups "inspired by, or associated with, designated foreign terrorist organizations or nations." They define domestic terrorists as individuals or groups aiming "to further ideological goals stemming from domestic influences, such as those of a political, religious, social, racial, or environmental nature."[2] However, there are prominent examples of U.S. domestic and foreign terrorist groups becoming more integrated to include having meetings and establishing networks in contested territories such as southeastern Ukraine (Rekawek 2020). There is also the example of the Russian Imperial Movement (RIM) that is fostering violent white supremacists worldwide

[2] *International terrorism*: Violent criminal acts committed by individuals and/or groups who are inspired by, or associated with, designated foreign terrorist organizations or nations (state-sponsored). (U.S. Code 2021), see https://www.govinfo.gov/content/pkg/USCODE-2009-title18/html/USCODE-2009-title18-partI-chap113B-sec2331.htm.

Domestic terrorism: Violent criminal acts committed by individuals and/or groups to further ideological goals stemming from domestic influences, such as those of a political, religious, social, racial, or environmental nature (Federal Bureau of Investigation 2023), see https://www.fbi.gov/file-repository/fbi-dhs-domestic-terrorism-definitions-terminology-methodology.pdf/view.

to include individuals within the United States.[3] If one of these groups with domestic and international counterparts were to acquire nuclear material, it would be particularly problematic for nuclear security.

RECOMMENDATION 3-1: The blurring of boundaries between state and non-state adversaries—such as the Wagner Group, Hamas, Hezbollah, and ISIS—raises the possibility that there may be gaps in U.S. government efforts to address nuclear threats. The committee recommends that the National Security Council and the Office of Management and Budget conduct a review of counterterrorism programs and agency budgets across the national security community to ensure that the attention being directed to Great Power Competition does not result in underinvesting in essential capabilities for managing and responding to the nuclear terrorism risk.

RECOMMENDATION 3-2: The transnational links among some anti-government/terrorist groups operating in the United States suggest that some of these groups might meet the criteria to be included on the list of Foreign Terrorist Organizations, which would make it illegal not only to join these groups but also to financially support them, as is the case for other listed FTOs. The Committee recommends the Departments of State, Treasury, Justice, Defense, and other relevant agencies examine these relationships and links to understand any such international connections and determine if any additional organizations are appropriate to add to this listing.

3.4 THE CONTINUUM BETWEEN STATE AND NON-STATE ACTORS

FINDING 3-3: Nuclear risks cannot be neatly divided between state and non-state actors. Instead, there is a continuum that blurs the line between state and non-state and challenges the distinction between acts of terrorism and acts of war. This gray zone includes state actors who use non-state actors as proxies. Proxies can present a "principal-agent" problem, in which the principal can lose control of the agent, as happened with Russia's Wagner Group in June 2023.

[3] In April 2020, the Department of State designated RIM, a Russian racially or ethnically motivated violent extremist (REMVE) organization suspected of attempting to incite violence in the United States, as a Specially Designated Global Terrorist (SDGT) organization. The threat from RIM—and from the Imperial Legion, RIM's military arm—lies not in the risk that RIM will attack the United States but in the reality that RIM is actively fostering violent white supremacist extremists worldwide (Gartenstein-Ross, Hodgson, and Clarke). The group runs paramilitary training camps in St. Petersburg and provided training to two Swedish men who placed three bombs (two detonating successfully) at a center for asylum seekers in Gothenburg, Sweden (Pompeo 2020). As documented by the Soufan Center in 2020, RIM has sent "foreign fighters" to conflicts in Ukraine, Syria, Libya, and the Central African Republic. The Soufan Center also documents a 2017 meeting between a RIM delegation and Matthew Heimbach, founder of the American neo-Nazi Traditionalist Worker Party, who helped organize the Unite the Right rally in Charlottesville, which led to the death of a counter protestor in 2017 (The Soufan Center 2020). The threat of attacks influenced by RIM in foreign countries is well established. A combination of existing evidence about Russia's malign influence efforts against the United States makes it clear that the threat to the Homeland is real and merits study and mitigation efforts.

ISIS is an example of a non-state actor that has attempted to hold territory and fulfill most of the functions of a state. Additionally, states continue to use proxy groups to conceal their involvement in domestic and international conflicts. For example, Pakistan's ISI has played an important role in the creation and support of the Taliban and extremist groups involved in Kashmir. Iran has used Hezbollah to support the Assad regime in Syria and in border disputes with Israel. Iran has also supported Hamas in its conflict with Israel. Both Hezbollah and Hamas also play political roles in Lebanon and Palestine, respectively, complicating their status still further.

A state's imperfect control of these proxies (terrorist, paramilitary, hacking, or mercenary group) makes it possible to claim that the agents are acting independent of the state. The advantage of this kind of arrangement is that it gives the state at least somewhat plausible deniability when attacks are successfully executed. The disadvantage is that the state may end up losing control of its agent. According to one well-established framework, states typically have three managerial strategies when it comes to proxy groups: (1) tacitly permitting (turning a blind eye), (2) orchestrating (actively funding or supporting a group in return for services), or (3) delegating (establishing command and control over the group). With each step toward "delegating," the state becomes more involved in the decisions and actions of the proxy group (Canfil 2022; Maurer 2018). The motivations of joining such groups vary. For example, in the case of an orchestrated proxy, the state and the proxy group are assumed to have similar interests, whereas, in the case of a mercenary group, the aim of the group is profit (McFate 2019).

Direct state complicity in a terrorist attack can have far-reaching diplomatic and political consequences to include elevating the intensity of a conflict. Accordingly, using a proxy can be attractive because it positions a state to attempt to shift blame, better mold its narrative, and create plausible, or in some cases even implausible, deniability surrounding an attack or mission. This method allows countries to advance their political or economic agenda over time and locations while minimizing the risk of accountability for their efforts and discouraging public attention (Atwell, Portzer, and McCurdy 2021). (That said, for reasons spelled out earlier, the risk of discovery would presumably largely deter states from covert delivery of nuclear weapons via proxies.)

According to a 2015 CSIS report on "Gray Zones," "Today, the toolkit for coercion below the level of direct warfare includes information operations, political coercion, economic coercion, cyber operations, proxy support, and provocation by state-controlled Forces. China, Russia, Iran, and North Korea, as well as non-state actors, are increasingly turning to these [gray-zone] strategies to overcome U.S. strengths in global diplomacy, law, and commerce" (Atwell, Portzer, and McCurdy 2021; Barno and Bensahel 2015).

The Wagner Group is a well-known example of a private group that operates in the grey area between state and non-state. Wagner is a paramilitary organization that has been trained and sponsored by the Russian Federation. Russia has been using the group to promote its foreign policy objectives, gather intelligence, influence political outcomes, and provide foreign militia training. Suspected and confirmed operations have occurred in as many as 30 countries and across four continents, including Ukraine, Syria, Libya, Central African Republic, Mali, and Venezuela (Katz et al. 2020). Notably, the Wagner Group, among other Russian private military groups, was used to destabilize and consolidate Russian power in Crimea and eastern Ukraine leading up to and supporting Russia's full-scale invasion in 2022. Russia's loss of control of the group in June 2023 illustrates the principal-agent problem that proxy groups can present (Rácz 2020).

3.5 THE LIMITS OF DETERRENCE IN PREVENTING NUCLEAR TERRORISM

Many nuclear weapon scenarios involving state sponsors, to include covert delivery via proxies, would be deterred if states assume their involvement could be attributed to the sponsoring state via nuclear forensics, or other intelligence, resulting in subsequent public identification and condemnation, and international retribution (National Academies of Sciences 2021). In short, deterrence works for rational state actors. Deterrence measures can also work for many non-state actors. While individual terrorists or suicide bombers may not be inhibited from employing a nuclear weapon or IND, their leaders would presumably want to survive to continue to pursue the terrorist groups' goals and maintain the public perception of their image and capabilities. Many terrorist groups could also be deterred by damage to their reputation, which might make them reluctant to violate nuclear norms.

Principles of deterrence are less likely to dissuade millenarian groups. This is because deterrence works best for adversaries that seek to maximize their chances of survivability and therefore seek to minimize the risk of massive retaliation. But many millenarian groups may actively court retaliation in the hope of sparking a broader war (Stern 2004).

In sum, when it comes to relying on deterrence for managing the nuclear terrorism threat, state-sponsored groups are more likely to be deterred than non-state-sponsored groups. However, it is also true that these kinds of non-state-sponsored groups are less likely to have access to nuclear weapons and materials. They are also unlikely to possess the expertise and operational capabilities to carry out a successful nuclear terrorist attack. Nonetheless, access and capability can change over time. If information becomes more available online, fissile and radiological materials proliferate without adequate security safeguards, and efforts to recruit insiders increase, more terrorist groups may be able to acquire the means to engage in nuclear terrorism.

3.6 NEW TYPES OF MILLENARIANISM AND TERRORIST RECRUITMENT OF MILITARY PERSONNEL

> FINDING 3-4: Examples of new types of millenarian groups—the type of terrorist group most likely to ignore anti-nuclear norms—are emerging. There is also increasing evidence that extreme-right-wing accelerationist groups are recruiting U.S. military personnel. If those personnel have insider knowledge about or access to nuclear facilities, materials, or intelligence, they may be in a position to compromise current U.S. nuclear security safeguards.

Millenarianism is the belief that after a major cataclysm, society will be dramatically changed or "cleansed." For secular millenarians, the kind of fundamental change they are pursuing arises by taking actions that have the potential to generate a political crisis to include sparking revolution. For example, there are contemporary domestic accelerationists who are actively seeking to incite a "second revolutionary war" in the United States. Since religious millenarians often believe that a savior or Messiah will appear following a period of tribulation, they seek to participate in bringing on the cataclysmic events that will cleanse the world in preparation for the Messiah. The threat of retaliation would presumably not deter these kinds of millenarian groups from extreme acts of violence to include nuclear or radiological devices, as they believe that cataclysmic events presage positive transformative change in society and in themselves. These are groups that want not a seat at the table but to blow up the table (Lemann 2001; National Commission on Terrorism 1998).

Examples of millenarian groups that have sought to acquire nuclear materials or weapons include not only al Qaeda and ISIS but also the group Atomwaffen. According to an investigation by Propublica, authorities discovered an aspirational plan to blow up a nuclear facility over 40 miles from Miami (Thompson, Winston, and Hanrahan 2018a). The group's name means "nuclear weapons" in German. Atomwaffen was formally disbanded in 2020 (Gais 2023), but it has evolved into a brand rather than a specific terrorist group (Lewis and Newhouse 2023; Shadnia et al. 2022). Members of the coalition associated with the Atomwaffen brand in the United States seek to incite a race war and overthrow the federal government.

Particularly troubling is Atomwaffen's recruitment of military or U.S. government personnel. Notably, the group's founder and leader, Brandon Russell, was arrested after authorities found homemade fuses, Geiger counters, and explosive and radiological materials in his garage. He had been serving in the 53rd Brigade Special Troops Battalion of Florida's Army National Guard at the time of his arrest (Goldwasser 2021). He was described by a former roommate as "obsessed with nuclear weapons" (Fleer 2020). After his 2021 release from prison, Russell was indicted, in 2023, for allegedly planning to disable the power grid in Maryland (Weiner et al. 2023). Josh Beckett, who served in the army from 2011 to 2015 as a combat engineer, trained Atomwaffen members in firearms and hand-to-hand combat and offered to construct weapons for the group (Thompson et al. 2018b). Naval aviation machinist mate's apprentice David Cole Tarkington was found to be a prolific Attomwaffen recruiter (Villarreal 2020).

The Base is another accelerationist neo-Nazi group operating in the United States, Europe, and Russia with the objective of establishing a white ethno-state. It is led by an American, Rinaldo Nazzaro, who fled to St. Petersburg, Russia, in 2020, where he still manages the group. Before founding The Base, Nazzarro was an analyst for the FBI and a contractor for the Department of Defense (DOD). He has also made unsubstantiated claims that he is a U.S. military veteran, an expert in defense studies, and a former CIA officer (Center for International Security and Cooperation 2021). Those with military training and experience with firearms and explosives are particularly valuable to the group. Its online application asked recruits about their training in the military, science, and engineering (The Southern Poverty Law Center 2022). In 2019, a member of the group told a federal prosecutor that Nazzaro was a Russian spy. The BBC has also reported that he was listed as a guest at a Russian government security exhibition that "focused on the demonstration of the results of state policy and achievements" (Arsenault and Stabile 2020).

The Order of 9 Angles (O9A) is a transnational accelerationist group that expresses admiration for both al Qaeda and Hitler. The O9A began as a British Satanic cult but is now a part of the international fascist revival. The O9A encourages its members to join military units. Several U.S. military personnel have been identified as members. One of them, Private Ethan Melzer, was sentenced to 45 years in March 2023 for releasing classified information, including about troop movements, to members of the group with the goal of facilitating a mass casualty attack on his own military unit (Levenson 2023).

FINDING 3-5: Domestic violent extremism is a potential threat to the security of the nuclear weapons enterprise and to the civil nuclear industry. The association of some U.S. government employees with domestic violent extremists provides them with additional capabilities that enhance their ability to carry out a successful attack.

By recruiting veterans and law enforcement agents, extremist groups are not only seeking technical and tactical expertise but also political legitimacy and mainstream credibility (Schake and Robinson 2021).The growing insider risk involving U.S. military personnel has been recognized by the leadership of the Department of Defense. An April 9, 2021, memorandum from defense secretary Lloyd Austin (U.S. Department of Defense 2021a) established a Countering Extremist Working Group for DOD that was directed to pursue four lines of effort: (1) military justice and policy, (2) support and oversight of the Insider Threat Program, (3) investigative processes and screening capability, and (4) education and training. The resulting "Report on Countering Extremist Activity within the Department of Defense" was released in December 2021 (U.S. Department of Defense 2021b) and consists of six recommendations and associated actions. The report identified the need to redefine the terms "extremist activities" and "active participation" found in department guidance. It also calls for updating service member transition checklists and coordination to support veterans in guarding against extremist recruiting. Additionally, it recommends the review and standardization of screening questionnaires. Finally, it calls for a commission to further study extremist activity that includes DOD civilians and contractors. All of these recommendations are well aligned with addressing the insider risk. But as of May 18, 2023, according to a Pentagon spokesperson, only the training recommendation has been implemented.

FINDING 3-6: The committee supports the work of the Countering Extremist Working Group and endorses DOD efforts to continue to enforce the identified measures for increasing awareness of domestic extremists. These efforts are relevant to all federal departments and agencies with employees and contractors who have access to sensitive information and facilities.

Concrete understanding and immediate identification of how Russia and other foreign actors exploit racially and ethnically motivated ideologies in online recruiting to radicalize vulnerable Americans and incite violence is critical to combating domestic terrorism and targeted violence. Ultimately, providing practitioners and policymakers involved in the work of designing prevention and intervention programs, including state, local, tribal, and territorial (SLTT) educational, health, and civic organizations, with a clear a foundational understanding of this threat is critical. Such efforts led by DHS should also include enhancing the ability of local law enforcement to identify any suspected terrorist threat, including nuclear threats.

Executive Order (E.O.) 13764 of January 17, 2017, is another important recognition of the importance of addressing the insider threat. This E.O. modified the existing civil service rules concerning security clearances, suitability, and fitness for employment, "to ensure that all persons performing work for or on behalf of the Government are and continue to be loyal to the United States" (The White House 2017).

For employees eligible for access to classified information, the E.O. directed that federal investigative standards "shall be designed to develop information as to whether the employment or retention in employment in the Federal service ... is consistent with the interests of national security." The new guidance requires that persons with access to classified information are subject to "continuous evaluation" and "continuous vetting." It explains that "'Vetting' is the process by which covered individuals undergo investigation, evaluation, and adjudication of whether they are, and remain over time, suitable or fit for Federal employment, eligible to occupy a sensitive position, eligible for access to classified information, eligible to serve as a nonappropriated fund employee or a contractor, eligible to serve in the military, or authorized to be issued a Federal credential." (The White House 2017) These continuous monitoring programs should be supplemented by dedicated programs to train and motivate staff with access to sensitive information and areas to identify and mitigate insider threats.

RECOMMENDATION 3-3: The U.S. government should maintain as a strategic priority the post-9/11 focus and effort on combating terrorism through ongoing deep collaboration and coordination across the national security community, in addition to international partners, SLTT authorities, the National Laboratories, universities and colleges, and civil society, and ensure that senior leaders at key agencies stay engaged in the counterterrorism mission.

RECOMMENDATION 3-4: To address the risk of radicalization by individuals who hold U.S. security clearances, the Administration should include the Department of Defense's revised definitions of "extremist activities" and "active participation," as described in the *Report on Countering Extremist Activity within the Department of Defense*, in the investigative standards for all government workers and contractors who have access to sensitive information and facilities as part of Executive Order 13764 of January 17, 2017, and as a part of the U.S. continuous vetting process.

References

Ackerman, Gary. 2023. "Nuclear Facilities Attack Database (NuFAD)." National Consortium for the Study of Terrorism and Responses to Terrorism (START). https://www.start.umd.edu/nuclear-facilities-attack-database-nufad.

Arsenault, Elizabeth Grimm, and Joseph Stabile. 2020. "Confronting Russia's Role in Transnational White Supremacist Extremism." *Just Security*. https://www.justsecurity.org/68420/confronting-russias-role-in-transnational-white-supremacist-extremism/.

Atwell, Kyle, Joshua M. Portzer, and Daphne McCurdy. 2021. "Negotiating [Im]plausible Deniability Strategic Guidelines for U.S. Engagement in Modern Indirect Warfare." *PRISM* 9 (2): 112–121. https://www.jstor.org/stable/27008979.

Barno, David, and Nora Bensahel. 2015. "Fighting and Winning in the 'Gray Zone.'" *War on the Rocks*. https://warontherocks.com/2015/05/fighting-and-winning-in-the-gray-zone/.

Benner, Katie, and Kate Conger. 2022. "U.S. Accuses 4 Russians of Hacking Infrastructure, Including Nuclear Plant." *New York Times*, March 24, 2022. https://www.nytimes.com/2022/03/24/us/politics/russians-cyberattacks-infrastructure-nuclear-plant.html.

Broad, William J. 2022. "Decades of Nuclear Reactor Strikes Predate Ukraine Power Plant Crisis." *New York Times*, October 9, 2022. https://www.nytimes.com/2022/10/09/science/ukraine-nuclear-power-plant-crisis.html.

Bunn, Matthew. 2016. "Belgium Highlights the Nuclear Terrorism Threat and Security Measures to Stop it." *Nuclear Security Matters*. https://www.belfercenter.org/publication/belgium-highlights-nuclear-terrorism-threat-and-security-measures-stop-it.

Bunn, Matthew, Martin B. Malin, Nickolas Roth, and William H. Tobey. March 2016. *Preventing Nuclear Terrorism: Continuous Improvement or Dangerous Decline?* Belfer Center for Science and International Affairs Harvard Kennedy School.

Canfil, Justin Key. 2022. "The Illogic of Plausible Deniability: Why Proxy Conflict in Cyberspace May No Longer Pay." *Journal of Cybersecurity* 8 (1). https://doi.org/10.1093/cybsec/tyac007.

Center for International Security and Cooperation. 2021. "The Base." Mapping Militant Organizations. https://cisac.fsi.stanford.edu/mappingmilitants/profiles/the-base#text_block_26733.

Council on Foreign Relations. 2006. "Loose Nukes." https://www.cfr.org/backgrounder/loose-nukes.

Earnhardt, Rebecca L., Brendan Hyatt, and Nickolas Roth. 2021. "A Threat to Confront: Far-Right Extremists and Nuclear Terrorism." *Bulletin of the Atomic Scientists*. https://thebulletin.org/2021/01/a-threat-to-confront-far-right-extremists-and-nuclear-terrorism/.

Federal Bureau of Investigation. 2023. "FBI: What We Investigate." Last Modified July 18, 2023. https://www.fbi.gov/investigate/terrorism.

Federation of American Scientists. 2023. "Nuclear Weapons: Nuclear Information Project." https://fas.org/issue/nuclear-weapons/.

Ferguson, Charles D., William C. Potter, Amy Sands, Leonard S. Spector, and Fred L. Wheling. 2005. *The Four Faces of Nuclear Terrorism*. Routledge.

Fleer, BreAnne K. 2020. "Radiological-Weapons Threats: Case Studies from the Extreme Right." *The Nonproliferation Review* 27 (1–3): 225–242. https://doi.org/10.1080/10736700.2020.1775987.

Gais, Hannah. 2023. "Atomwaffen Co-Founder Pleads Guilty to Double Murder Charges." *Hatewatch*. https://www.splcenter.org/hatewatch/2023/05/11/atomwaffen-co-founder-pleads-guilty-double-murder-charges.

Gartenstein-Ross, Daveed, Samuel Hodgson, and Colin P. Clarke. 2020. "The Russian Imperial Movement (RIM) and Its Links to the Transnational White Supremacist Extremist Movement." *The International Centre for Counter-Terrorism*. https://www.icct.nl/publication/russian-imperial-movement-rim-and-its-links-transnational-white-supremacist-extremist.

Goldwasser, Rachel. 2021. "Extremism among Active-Duty Military and Veterans Remains a Clear and Present Danger." *Hatewatch*. https://www.splcenter.org/hatewatch/2021/10/12/extremism-among-active-duty-military-and-veterans-remains-clear-and-present-danger.

Institute for Economics & Peace. 2023. *Global Peace Index*. https://www.visionofhumanity.org/wp-content/uploads/2023/06/GPI-2023-Web.pdf

Katz, Brian, Seth G. Jones, Catrina Doxsee, and Nicholas Harrington. 2020. "Moscow's Mercenary Wars. The Expansion of Russian Private Military Companies." https://russianpmcs.csis.org/.

Kuleba, Dymtro (@DmytroKuleba). 2022. "The only electrical grid supplying the Chornobyl NPP and all its nuclear facilities occupied by Russian army is damaged. CNPP lost all electric supply. I call on the international community to urgently demand Russia to cease fire and allow repair units to restore power supply 1/2." Twitter, March 9, 2022, 6:11 AM. https://twitter.com/DmytroKuleba/status/1501531157510426625.

Lemann, Nicholas. 2001. "What Terrorists Want." *New Yorker*, October 21, 2001. https://www.newyorker.com/magazine/2001/10/29/what-terrorists-want.

Levenson, Michael. 2023. "Former Army Private Sentenced to 45 Years in Neo-Nazi Plot to Kill Soldiers." *New York Times*, March 3, 2023. https://www.nytimes.com/2023/03/03/us/ethan-melzer-army-neo-nazi-sentenced.html.

Lewis, Jon, and Alex Newhouse. 2023. "Be Careful Attributing Anything to AWD." https://www.accresearch.org/shortanalysis/be-careful-attributing-anything-to-awd.

Maurer, Tim. 2018. *Cyber Mercenaries: The State, Hackers, and Power*. Cambridge University Press.

McFate, Sean. 2019. *Mercenaries and War: Understanding Private Armies Today*. Washington, DC: National Defense University Press.

National Academies of Sciences, Engineering, and Medicine. 2021. *Restoring and Improving Nuclear Forensics to Support Attribution and Deterrence: Public Summary*. Washington, DC: The National Academies Press.

National Commission on Terrorism. 1998. "Countering the Changing Threat of International Terrorism."

Nuclear News. 2022. "Indictment Related to Wolf Creek Computer Hack Unsealed." *Nuclear Newswire*. https://www.ans.org/news/article-3818/indictment-related-to-wolf-creek-computer-hack-unsealed/.

Pompeo, Michael R. 2020. "United States Designates Russian Imperial Movement and Leaders as Global Terrorists." U.S. Department of State press release. https://2017-2021.state.gov/united-states-designates-russian-imperial-movement-and-leaders-as-global-terrorists/.

Rácz, András. 2020. "Band of Brothers: The Wagner Group and the Russian State." Understanding the Russian Military Today. https://www.csis.org/blogs/post-soviet-post/band-brothers-wagner-group-and-russian-state.

Rekawek, Kacper. 2020. *Career Break or a New Career? Extremist Foreign Fighters in Ukraine*. Germany: Counter Extremism Project.

Ritter, Karl, and Jon Gambrell. 2022. "Ukraine Says a Russian Missile Struck Close to a Nuclear Power Plant." *Associated Press*, September 19, 2022. https://www.npr.org/2022/09/19/1123782942/ukraine-russian-missile-strike-near-nuclear-power-plant.

Roza, David. 2022. "The US Military Is Still Missing 6 Nuclear Weapons That Were Lost Decades Ago." https://taskandpurpose.com/history/us-military-nuclear-weapons-missing/.

Schake, Kori, and Michael Robinson. 2021. "The Military's Extremism Problem Is Our Problem." *New York Times*, March 2, 2021. https://www.aei.org/op-eds/the-militarys-extremism-problem-is-our-problem/.

Shadnia, Deeba, Alex Newhouse, Matt Kriner, and Arthur Bradley. 2022. "Militant Accelerationism Coalitions: A Case Study in Neo-Fascist Accelerationist Coalition Building Online." *Tech against Terrorism*. https://www.middlebury.edu/institute/sites/www.middlebury.edu.institute/files/2022-06/REDACTED%20CTEC__TAT%20Accelerationism%20Report%20.pdf?fv=zvPnbbzC.

Stern, Jessica. 2004. *Terror in the Name of God: Why Religious Militants Kill*. Ecco.

Suciu, Peter. 2021. "The U.S. Military Is Missing Six Nuclear Weapons." *National Interest*. https://nationalinterest.org/blog/reboot/us-military-missing-six-nuclear-weapons-180032.

Swan, Betsy Woodruff, and Maggie Miller. 2022. "Russian Spies Indicted in Worldwide Hacks of Energy Industry, Including Kansas Nuclear Plant." *Politico*. https://www.politico.com/news/2022/03/24/russian-spies-indicted-hack-nuclear-plant-00020217.

The Soufan Center. 2020. "Inside the Russian Imperial Movement: Practical Implications of U.S. Sanctions." https://thesoufancenter.org/wp-content/uploads/2020/06/TSC-Report-Inside-the-Russian-Imperial-Movement-Practical-Implications-of-U.S.-Sanctions.pdf.

The Southern Poverty Law Center. 2022. "The Base." *Hatewatch*. https://www.splcenter.org/fighting-hate/extremist-files/group/base.

The White House. 2017. *Amending the Civil Service Rules, Executive Order 13488, and Executive Order 13467 to Modernize the Executive Branch-Wide Governance Structure and Processes for Security Clearances, Suitability and Fitness for Employment, and Credentialing, and Related Matters*. Federal Register.

Thompson, A. C., Ali Winston, and Jake Hanrahan. 2018a. "Inside Atomwaffen as It Celebrates a Member for Allegedly Killing a Gay Jewish College Student." *ProPublica*, February 23, 2018. https://www.propublica.org/article/atomwaffen-division-inside-white-hate-group.

---. 2018b. "Ranks of Notorious Hate Group Include Active-Duty Military." *ProPublica*, May 3, 2018. https://www.propublica.org/article/atomwaffen-division-hate-group-active-duty-military.

U.S. Code. 2021. United States Code, 2018 Edition, Supplement 3, Title 18 - Crimes and Criminal Procedure.

U.S. Department of Defense. 2021a. *Immediate Actions to Counter Extremism in the Military and the Establishment of the Countering Extremism Working Group*.

---. 2021b. *Report on Countering Extremist Activity within the Department of Defense*. https://media.defense.gov/2021/Dec/20/2002912573/-1/-1/0/report-on-countering-extremist-activity-within-the-department-of-defense.pdf.

Villarreal, Daniel. 2020. "Navy Kicks Out Alleged Recruiter for Neo-Nazi Group Atomwaffen Division after Investigation." *Newsweek*, April 17, 2020. https://www.newsweek.com/navy-kicks-out-alleged-recruiter-neo-nazi-group-atomwaffen-division-after-investigation-1498704.

Ward, Antonia. 2018. "Is the Threat of Nuclear Terrorism Distracting Attention from More Realistic Threats?" https://www.rand.org/blog/2018/07/is-the-threat-of-nuclear-terrorism-distracting-attention.html.

Weiner, Rachel, Jasmine Hilton, and Dan Morse. 2023. "Duo Accused of Neo-Nazi Plot to Target Maryland Power Stations." *Washington Post*, February 6, 2023. https://www.washingtonpost.com/dc-md-va/2023/02/06/maryland-power-grid-neonazi-brandon-russell/.

Partner Nations in the Global Initiative to Combat Nuclear Terrorism

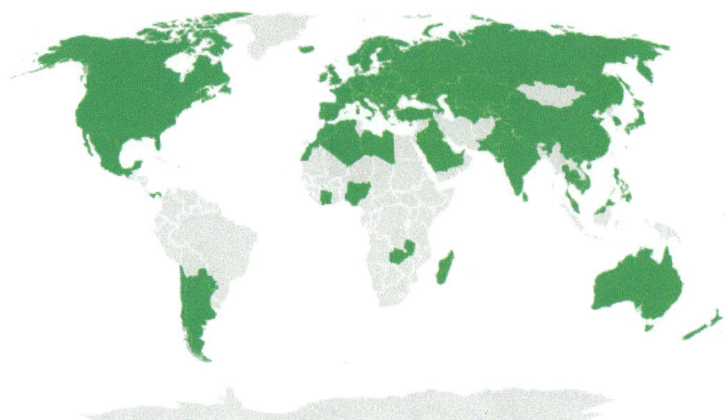

FIGURE 4-1 The Global Initiative to Combat Nuclear Terrorism (GICNT), a voluntary partnership of 89 nations and six international organizations with a mission to prevent, detect, and respond to nuclear terrorism, has multilateral activities that strengthen the plans, policies, procedures, and interoperability of partner nations. The United States and the Russian Federation serve as co-chairs, while Morocco currently leads the Implementation and Assessment Group (IAG) under the guidance of the co-chairs. GICNT activities were paused after Ukraine was invaded by the Russian state, an example of how new geopolitical instability is challenging the U.S. efforts to prevent and counter nuclear terrorism.
SOURCE: U.S. Department of State 2017.

4

Geopolitical and Other Changes Eroding Long-Standing Nuclear Security Norms and Practices

4.1 BACKGROUND

Since the development of the first atomic weapon and the bombings of Hiroshima and Nagasaki at the end of World War II, there has been global concern about how to limit the use and spread of nuclear weapons. Throughout the Cold War, while the United States and the former Soviet Union built up massive nuclear arsenals, they also reached agreements to stabilize the numbers of these weapons and reduce the risk of inadvertently crossing the nuclear threshold. The fall of the Soviet Union in 1991 brought with it fear that the Russian government and the newly independent states of Ukraine, Belarus, and Kazakhstan would be unable to effectively control the nuclear weapons within their jurisdictions. This led to joint U.S. and Russian programs to protect and repatriate these weapons, and to protect Russian nuclear material stockpiles made vulnerable by the abrupt disappearance of the centralized command-and-control system.

A decade later, the 9/11 terrorist attacks again elevated concerns about the threat of nuclear terrorism by non-state actors. The dissolution of the Soviet Union along with the threat of global terrorism led to the development of a range of U.S. and other cooperative international government-to-government programs, as well as new international frameworks and agreements, to include:

- The Department of Defense (DOD) Cooperative Threat Reduction (CTR) program, which has evolved into Material Management and Minimization in the Department of Energy (DOE), and multiple program efforts managed by the Defense Threat Reduction Agency (DTRA) in the DOD.
- Material Protection, Control, & Accounting (MPC&A), a DOE program that also grew out of CTR and is currently the International Nuclear Security Program (INS).

BOX 4-1
Summary

A unifying theme in this report is the indispensable role that the United States has played and must continue to play in mobilizing and sustaining global efforts to advance nuclear security. Renewed attention to this imperative is especially important given the erosion in recent years of many of the post–Cold War norms that supported international cooperation in addressing the risk of nuclear terrorism. Most prominent among this shift is Russia's transition from an important partner in enhancing nuclear security to a violator of long-standing nuclear taboos. This includes the Russian threat to use nuclear weapons as a coercive tool in its war on Ukraine, which had chosen to become a non-nuclear state when the Soviet Union dissolved. This undermines nonproliferation efforts by demonstrating the possible usefulness for an aggressor of possessing nuclear weapons and the relative weakness of a state that does not (Committee Meeting, November 29, 2022).[a] Given this new reality, there must be strong U.S.-led efforts to adapt and expand the international programs that have to date prevented a successful terrorist nuclear attack and have slowed the proliferation of nuclear weapons.

For three decades, the cornerstone of managing the nuclear terrorism threat has been limiting the number of nuclear weapons and the availability of weapons-usable nuclear materials that may potentially fall into the hands of non-state actors. In recent years, global partnerships in support of arms control, nonproliferation, and combating nuclear terrorism have weakened. The Nuclear Security Summit process that mobilized and focused international attention on the need to manage the risk of nuclear weapons-usable materials ended in 2016 (Bunn 2016; Gill 2020). It was followed by a U.S. administration that, although it made statements in support of international institutions to reduce nuclear terrorism threats, espoused doubts about the effectiveness of international obligations and responsibilities (Arms Control Association 2019; Roth 2020). Meanwhile, China, Russia, India, Pakistan, and North Korea continue to expand their nuclear weapons programs, fueling the anxieties of

- The Second Line of Defense, which included the Megaports Initiative and is now Nuclear Smuggling Detection and Deterrence (NSDD) in the DOE.
- The Global Threat Reduction Initiative (GTRI), now the Radiological Security Program in the DOE.
- The Container Security Initiative (CSI) in the DHS.
- The Customs Trade Partnership against Terrorism (C-TPAT) in the Department of Homeland Security (DHS).
- The Proliferation Security Initiative (PSI).
- International Ship and Port Facility Security Code (ISPS).

other countries in Asia and Europe. Iran has begun to enrich uranium to 60% and above in violation of its obligations under the JCPOA, which could stimulate interest in enrichment in other countries in the Middle East (Cordesman 2021; International Atomic Energy Agency 2023; Lerner 2022; Murphy 2023). Fortunately, intergovernmental and international organizations, such as the International Atomic Energy Agency (IAEA), have expanded their focus on the security of nuclear materials.

Given this dynamic threat environment, there is an urgent need for the United States to reinvigorate efforts to engage heads of state to work together to close any existing or emerging gaps in the international nuclear security system. Additionally, U.S. prevention and protection programs carried out in cooperation with organizations such as the IAEA and Interpol, as well as with like-minded countries, require increased funding and coordination. A more integrated international approach is needed for managing the evolving nuclear terrorism risk.

Highlights

- Destabilizing changes in the political, societal, and technological environment over the last several years require an increased focus on preventing nuclear and radiological terrorism.
- The increasing amount of nuclear and radiological material being produced worldwide could elevate the amount of material at risk if strict security and materials safeguards are not in place.

Without dialogue between the United States and Russia, and a significant change in the current U.S. relationship with China, cooperation on nuclear and terrorism issues among the three largest nuclear states will vanish.

[a] Written materials submitted to a study committee by external sources and public meeting recordings are listed in the project's Public Access File and can be made available to the public upon request. Contact the Public Access Records Office (PARO) at the National Academies of Sciences, Engineering, and Medicine for a copy of the list and to obtain copies of the materials. E-mail: paro@nas.edu.

- U.N. Security Council Resolution 1540 (UNSCR 1540).
- Nuclear Forensics International Technical Working Group (ITWG).
- Treaty on the Non-Proliferation of Nuclear Weapons (NPT).
- International Convention for the Suppression of Acts of Nuclear Terrorism (ICSANT).
- The Global Initiative to Combat Nuclear Terrorism (GICNT).

For an NNSA overview of these strategies and U.S. policies, refer to the NNSA strategic plan (National Nuclear Security Administration 2021).

4.2 WEAKENING POLITICAL, SOCIETAL, AND TECHNOLOGICAL ENVIRONMENT

> FINDING 4-1: Destabilizing changes in the political, societal, and technological environment over the last several years require an increased focus on preventing nuclear terrorism, both by strengthening and expanding government prevention programs and by reinvigorating and extending international cooperation in this area. These adverse changes include: the increasing amount of nuclear and radiological material being produced worldwide (see chapters 6 and 7); the erosion of norms and international collaboration around nuclear threats and nonproliferation (chapter 4); the blurring of the distinction between state and non-state actors (chapter 3); persistent insider threats (chapter 3); and powerful new technologies for sharing information and misinformation, for manufacturing weapons, and for weapon delivery (discussed in the classified annex to this report).

While it is difficult to measure, verify, and validate the individual contribution to nuclear security of each of these programs and agreements listed above, the historical record is clear: to date no terrorist group has detonated a nuclear weapon, an improvised nuclear device (IND), or a radiological dispersal device (RDD). However, the absence of catastrophe or limited knowledge of major security incidents can contribute to complacency, especially when there are many other urgent issues competing for the attention of national leaders as well as budgetary challenges. As the world moves toward the eighth decade of the nuclear era, there are worrisome indications that the nuclear threat, and in particular the threat from non-state actors, is not receiving the active and ongoing attention it once had by political leaders and the general public.

When it comes to the threat posed by non-state actors, the foundations of nuclear security are well established. The greatest impediment to making an IND or RDD has always been the challenge of obtaining a sufficient quantity of nuclear or radiological material. Accordingly, prevention programs have been designed to make such efforts by non-state actors prohibitively difficult. These programs pursue three outcomes: (1) reducing the amount of nuclear and radiological material available; (2) effectively protecting the material that remains; and (3) detecting and intercepting any material that has moved out of regulatory control. Significant achievements by the United States and its international partners have been made in all three of these areas. However, there are new developments that will most likely lead to a significant increase in nuclear material to include the expansion of arsenals by current nuclear weapons states, the likely emergence of new nuclear weapons states, and the renewed global interest in civil nuclear power as a sustainable source of energy.

A new chapter in the nuclear era, fraught with risks and challenges but also opportunities, is unfolding. Many countries are embracing the development of new nuclear power programs despite lacking any experience in managing and safeguarding nuclear material and facilities. In the 1990s, Ukraine had returned to Russia the Soviet-era nuclear weapons that had been stationed there, under Soviet operational control. As a consequence of this and long-standing insecurities about reliance on allies, some countries may think obtaining nuclear weapons is in their interest and others may be reluctant to relinquish any special nuclear material already in their possession. Still others may be motivated to manufacture, or develop the capacity to manufacture, nuclear materials.

The sobering current reality is that the international institutions and partnerships that have been the cornerstone of nonproliferation and counterproliferation efforts are faltering. The disruptive role Russia is currently playing on the world stage is a growing challenge, given that it possesses the world's largest nuclear arsenal. Additionally, high-level political attention among national leaders has waned since 2016 with the end of the Nuclear Security Summit process initiated by President Barack Obama in 2010. The Global Initiative to Combat Nuclear Terrorism, jointly led by the United States and Russia, has paused its meetings and working groups. China and North Korea, along with Russia, are expanding their nuclear programs. Pakistan faces challenges to its stability and is locked in an ongoing conflict with India, another nuclear state; Iran continues to produce weapons-usable nuclear materials. Finally, quantities of separated plutonium in the civilian sector are increasing in many countries.

New technologies are also introducing new challenges for nuclear security. While obtaining material continues to be a significant hurdle to developing an IND or RDD, powerful communication tools have made obtaining information about the location of material and how to weaponize it increasingly available, although the accuracy is suspect. New and accessible technologies—such as AI, additive manufacturing, and drones—may simplify the manufacturing and delivery process as well as create new threats and vulnerabilities at nuclear facilities. These trends all point to the need for urgency in bolstering efforts to reduce and protect material, to develop countermeasures for new attack vectors, and to develop a rejuvenated international network of countries working to counter proliferation.

The risks and challenges posed by the expanding and evolving civil nuclear energy sector and those posed by the most dangerous nuclear materials, highly enriched uranium (HEU) and separated plutonium, will be addressed at greater length elsewhere in this report. So too will the issues associated with radiological materials that can be used to fabricate an RDD, given the amount and accessibility of these materials around the world.

4.3 THE ERA OF GREAT POWER COMPETITION AND COUNTERING NUCLEAR TERRORISM

FINDING 4-2: The broken U.S. relationship with Russia, and the increasingly distrustful U.S. relationship with China, have led to a major reduction in cooperation on nuclear and terrorism issues and greater uncertainty about security of the world's largest nuclear stockpiles.

The once robust collaboration with Russia to secure its nuclear weapons and nuclear weapons-usable fissile materials, and the repatriation of weapons and material from the newly independent states, was the origin for many of the important U.S. government and international programs to prevent nuclear terrorism. These cooperative efforts also eventually expanded to include work to jointly develop and deploy detection equipment at key Russian land and sea borders with the aim of deterring and intercepting the possible smuggling of uncontrolled nuclear weapons and material. With the post-9/11 focus on preventing terrorists from obtaining nuclear weapons and material, these programs were all expanded to countries outside Russia.

The extensive experience developed through the Russian material protection control and accounting work is currently reflected in the DOE International Nuclear Security Program, which works at both the national and site level with long-term and emerging nuclear states. The post-Soviet removal of nuclear material from the newly independent states by the DOD and DOE expanded into a major worldwide effort by the DOE during the Nuclear Security Summit era. It is now being managed by the DOE/NNSA Office of Material Management and Minimization. Early efforts to secure radiological sources were integrated into a single DOE program that is focused on the consolidation and protection of radiological material both domestically and internationally. Finally, the collaborative work to deploy detection equipment along Russian borders evolved into the DOE Nuclear Smuggling Detection and Deterrence Program, which is providing equipment and working with customs, border police, and investigative agencies along land and sea borders worldwide.

All of these programs would benefit from additional investments, enhanced integration and coordination, and renewed high-level support. Specific recommendations for a subset of these and related programs are outlined in chapter 8.

While these and additional important programs led by other U.S. agencies continue to do excellent work, the collaboration between Russia and the United States, which underpinned many of the global nonproliferation, counterproliferation, and counterterrorism efforts, is no longer operative since Russia's first invasion of Ukraine. An additional example of the collapsed cooperative relationship is Russia's February 2023 announcement that it will suspend participation in New Start (U.S. Department of State 2023), the last remaining strategic arms control treaty with the United States.

The history of U.S. cooperation with China on nuclear- and terrorism-related issues was never as advanced as U.S. cooperation with Russia. More recently, the increasingly intense competition and distrust between the United States and China has resulted in a significantly reduced level of cooperation in all areas, at a time when China is undertaking a significant expansion of its nuclear weapons program.

The March 2023 release of National Security Memorandum 19 (NSM-19) provides an updated U.S. government strategy to "Counter Mass Destruction Terrorism and Advance Nuclear and Radioactive Material Security." NSM-19 clarifies the overall strategic framework along three mission areas (The White House 2023):

- Counter WMD terrorism: "Combatting all stages of WMD terrorism requires constant vigilance against an ever-changing threat landscape. This NSM also serves as a call to action to disrupt and hold accountable those who provide material, financial, or other support to non-state actors seeking WMD capabilities."

- Advance nuclear material security: "The peaceful uses of nuclear technology provide considerable economic, medical, and environmental benefits. However, the storage, transportation, processing, and use of highly enriched uranium, separated plutonium, and other weapons-usable nuclear material globally present persistent national security risks to the United States."
- Advance radioactive material security: "As with nuclear materials and technology, the peaceful uses of radioactive materials provide considerable benefits, although the storage, transportation, processing, and use of radioactive materials globally present a security risk that must be addressed through collective and continuous efforts. Minimizing the use of these materials where technically and economically feasible alternatives exist reduces risk in our collective national security, health, and economic interest."

RECOMMENDATION 4-1: Based on the Biden Administration's recently released *Strategy for Countering Weapons of Mass Destruction Terrorism* (NSM-19), the U.S. government, led by the National Security Council, should continue to prioritize and provide oversight of a "whole of government"/ "whole of nation"[1] focus on preventing nuclear terrorism, to include strengthening and extending ongoing nonproliferation and counterproliferation programs.

Under this strategic direction, agencies such as the DOE, DOD, and the state, with one-of-a-kind programs and unique capabilities, would proactively continue to identify gaps, overlaps, and new opportunities to respond to the changing threat environment. Congress should adequately fund these programs and activities as well as the federal workforce and staff, and provide regular oversight to include conducting hearings.

RECOMMENDATION 4-2: Combating the threat of nuclear terrorism is a shared global interest; the U.S. government should provide strong and visible international leadership as it has done in the past.

Strong U.S. leadership would influence and encourage collaboration among global leaders to prioritize domestic, bilateral, and multilateral nuclear security activities. Such collaboration and cooperation would include capacity-building to prepare for and respond to the changing risk environment. The full range of diplomatic efforts should be brought to bear on the problem and should include engagements with governments that have the resources and technical know-how to contribute to these efforts. Moreover, as appropriate and politically feasible, these efforts should include Russia and China as well as other states with which the United States and its allies have difficult and complicated relations.

---

[1] In the report, the committee stresses the importance of coordination and capacity-building with state, local, tribal, and territorial (SLTT) governments. The addition of "whole of nation" was added as a reminder that this is not singularly an interagency activity at the federal level but extends to SLTT involvement.

References

Arms Control Association. 2019. "A Critical Evaluation of the Trump Administration's Nuclear Weapons Policies." *Arms Control Today*. https://www.armscontrol.org/events/2019-07/critical-evaluation-trump-administrations-nuclear-weapons-policies.

Bunn, Matthew. 2016. "The Nuclear Security Summit: Wins, Losses, and Draws." *All Nuclear Security Matters*. https://dash.harvard.edu/handle/1/37375266.

Cordesman, Anthony H. June 30, 2021. *Iran and U.S. Strategy: Looking beyond the JCPOA*. Center for Strategic and International Studies (CSIS). http://www.jstor.org/stable/resrep33760.

Gill, Amandeep S. 2020. *Nuclear Security Summits A History*. Cham: Palgrave Macmillan.

International Atomic Energy Agency. 2023. "Verification and Monitoring in the Islamic Republic of Iran in Light of United Nations Security Council Resolution 2231 (2015)." GOV/2023/24. https://www.iaea.org/sites/default/files/23/06/gov2023-24.pdf.

Lerner, K. Lee. 2022. "Policymakers Must Now Assume That Iran Has the Enriched Uranium It Needs to Build a Nuclear Weapon." *Taking Bearings*. Harvard Blogs (June 1). https://blogs.harvard.edu/kleelerner/iran-now-has-the-enriched-uranium-it-needs-to-build-a-nuclear-weapon/.

Murphy, Francois. 2023. "Iran Expands Stock of Near-Weapons Grade Uranium, IAEA Reports No Progress." *Reuters*, September 4, 2023. https://www.reuters.com/world/middle-east/iaea-reports-no-progress-iran-uranium-stock-enriched-60-grows-2023-09-04/.

National Nuclear Security Administration. 2021. "Prevent, Counter, and Respond—A Strategic Plan to Reduce Global Nuclear Threats (NPCR)." https://www.energy.gov/nnsa/articles/prevent-counter-and-respond-strategic-plan-reduce-global-nuclear-threats-npcr.

Roth, Nickolas. 2020. "Preventing Nuclear Terrorism." *The Trump Administration after Three Years: Blundering toward Nuclear Chaos,* edited by Jan Wolfsthal. American Nuclear Policy Initiative. https://www.globalzero.org/wp-content/uploads/2020/05/Blundering-Toward-Nuclear-Chaos-ANPI-Report-May-2020.pdf.

The White House. 2023. "Fact Sheet: President Biden Signs National Security Memorandum to Counter Weapons of Mass Destruction Terrorism and Advance Nuclear and Radioactive Material Security". https://www.whitehouse.gov/briefing-room/statements-releases/2023/03/02/fact-sheet-president-biden-signs-national-security-memorandum-to-counter-weapons-of-mass-destruction-terrorism-and-advance-nuclear-and-radioactive-material-security/

U.S. Department of State. 2017. *Global Initiative to Combat Nuclear Terrorism*.

---. 2023. *Russian Noncompliance with and Invalid Suspension of the New START Treaty*.

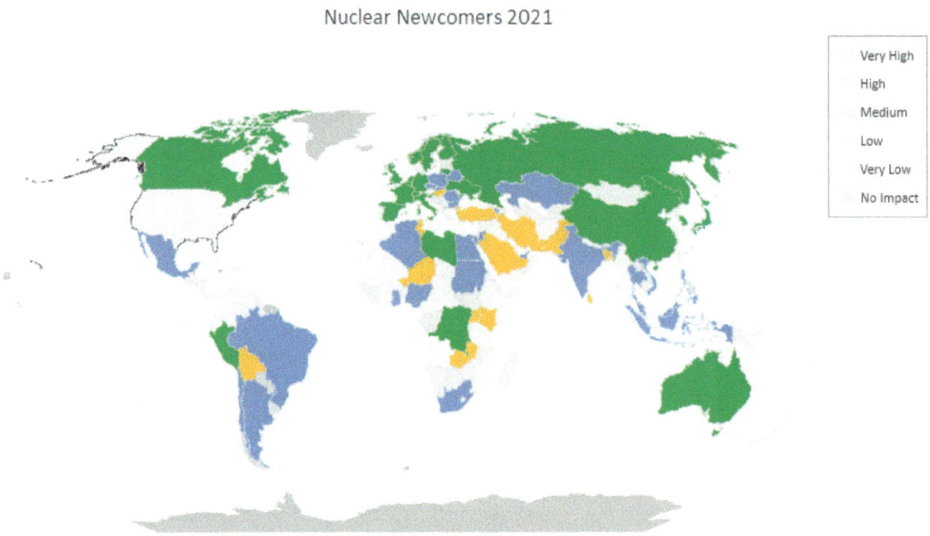

FIGURE 5-1 A global map showing the countries currently considering or embarking on nuclear power as a dependable, low-carbon energy source, shown in blue and yellow. These countries do not have previous experience with nuclear energy. The countries in blue are partner countries with the International Nuclear Security (INS) program at the National Nuclear Security Administration; yellow countries have not partnered with INS. The countries in green are INS partners and have nuclear power programs.
SOURCE: Holt 2021.

5

The Evolving Civil Nuclear Energy Sector: Adapting Approaches and New Opportunities

5.1 EXPANDING CIVIL NUCLEAR ENERGY SECTOR SECURITY

Currently, nuclear reactors provide 20 percent of the electrical generation in the United States and are the largest single contributor to clean energy generation (over 70 percent) (U.S. Department of Energy Office of Nuclear Energy [DOE-NE] 2023). As new sources of clean electricity generation are needed to meet expected future increases in electricity demand, new nuclear reactors and new types of nuclear reactors can play a role in meeting this new demand. These reactors will no doubt include small modular reactors (SMR), many of which will be derivations of light-water reactor technology, as well as new types of advanced reactors based on new fuel concepts and with substantially different designs from those operated commercially today. Some of these new reactors may also provide heat and power for non-electric applications (DOE-NE 2023). For example, advanced technology reactors may be uniquely suited for water desalination and cogeneration alongside various industrial processes, process heat, and hydrogen generation, and could be developed in smaller, truck-portable variations (Nuclear Energy Institute 2022).

According to the International Energy Agency, nuclear power's global contribution would need to double from 413 GW in early 2022 to 812 GW in 2050 to achieve desired net-zero emissions scenarios. To achieve this level of production would require annual nuclear capacity additions averaging 27 GW per year in the 2030s, higher than any decade before. Emerging and developing economies would account for more than 90 percent of global growth, with China set to become the leading nuclear power producer before 2030 (International Energy Agency 2022). This would translate into new reactors being deployed in countries that have no history or experience with nuclear power or nuclear reactors, and thus no or limited experience with nuclear safety and safeguards.

BOX 5-1
Summary

Looking ahead there will be expanding opportunities for non-state actors to obtain access to nuclear materials or to target nuclear facilities. This is due to the projected increase in the number of civil nuclear facilities and the volume of nuclear fresh and spent fuel in transit. More civil nuclear material and nuclear facilities around the globe will require a strategy to ensure their security from terrorist attacks and to counter proliferation.

International interest in civilian nuclear energy is growing due to its potential to provide clean power and support the goal of achieving net-zero carbon emissions. At the same time, there are new safety and security considerations as nuclear power becomes more accessible due to unique designs and technological improvements inherent in small modular reactors and other advanced reactors. Led primarily by non-U.S. corporations, the civil nuclear energy sector is now expanding into countries that lack experience with nuclear safety and safeguards, so-called nuclear newcomers. Meanwhile, Russian attacks on Ukrainian nuclear power plants and other energy supply and distribution systems have for the first time introduced the possibility that an operating civil nuclear power plant could be targeted by state and non-state actors to terrorize residents and persuade them to bend to the will of the attacker.

Keeping spent nuclear fuel in pools on the sites of nuclear power plants elevates the risk of radiation release as a result of a terrorist attack or military action at or near a facility. If most spent fuel is stored in licensed hardened storage containers, rather than in fuel pools, it will be less vulnerable. But permanent solutions for disposal of spent nuclear fuel must be found, including in the United States. This is the most practical alternative for ensuring that spent nuclear fuel is adequately safeguarded for extended periods of time.

The U.S. nuclear industry dominated the global market for nuclear power export throughout the 1970s and 1980s, thus collaterally exporting exceedingly high standards for nuclear safety, security, and nonproliferation. Without U.S. leadership during the upcoming wave of deployment, assurance that new entrant reactor vendors

Some of the new technologies could use fuel with higher uranium enrichments than those commonly in use today. Although varying in design, many advanced reactor fuels would require what is termed "high assay low enriched uranium" or HALEU, using uranium enriched up to 19.95 percent. A category of advanced reactors, termed "fast" reactors, would be designed to breed more fissile material than they consume. Additional new fuel types could use plutonium, including surplus weapons-grade plutonium. These new fuel and reactor types, some of which could be more attractive terrorist targets compared to existing targets, will need new approaches to safeguards and security, particularly those with new deployment concepts, such as mobile, sea-based, and remote sites.

and suppliers will adopt similarly high standards may be lost. Strong U.S. leadership and presence in global markets is essential as nuclear energy plays a larger role in clean energy transitions around the globe. This includes forging a transparent and productive partnership among the U.S. government, the nuclear industry, and the International Atomic Energy Agency in establishing the export and adoption of high standards of safety, security, and safeguards.

Nuclear security is not as universally formalized and instituted as are nuclear safety and nuclear safeguards. The participation by U.S. government and private sector experts in international, multilateral initiatives such as the International Atomic Energy Agency Nuclear Harmonization and Standardization Initiative has made a positive contribution toward achieving the goal of safe and secure deployment of small modular reactors and other advanced nuclear technologies, while maximizing the potential contribution of such technologies to achieve global clean energy goals. The United States, however, needs to move beyond participation and both lead and drive international standards-setting and regulatory harmonization efforts for attaining high standards and norms around nonproliferation, materials control and accounting, and physical and cybersecurity for these advanced nuclear technologies.

Highlights

- As a result of increasing reliance on nuclear power, there will be more nuclear material and facilities around the globe. This requires new ways of thinking about security particularly in nuclear newcomer states.
- Like most energy infrastructure, civil nuclear energy facilities are not hardened against a nation-state assault under military conditions of sustained war fighting as seen in Ukraine at the Zaporizhzhia nuclear power plant.
- The U.S. is working to regain leadership in nonproliferation, materials control and accounting, and physical and cybersecurity for advanced nuclear technologies.

With adequate resources, the IAEA can play an indispensable role in strengthening international standards on nuclear safety, security, and material control and accounting for advanced nuclear technologies.

According to a global deployment analysis by Third Way, between 2021 and 2050, total electricity consumption is projected to approximately double to over 50,000 terawatt hours annually (Ahn and Allen 2022). Around 75 percent of new demand will come outside of developed countries. A total of 52 countries are projected to be markets for advanced nuclear power before 2050. Nine countries that do not currently have operating commercial nuclear plants are considered viable markets today, and an additional 10 countries are anticipated to have advanced nuclear reactors by 2030, and likely would be by 2050 (Ahn and Allen 2022). The global market for nuclear power could potentially triple by 2050. Additional demand drivers such as water desalination, industrial heat, electric vehicles, and coal replacement could increase projected new nuclear generation and market size even further.

The Department of Energy received $230 million in fiscal year 2020 to start a program for an Advanced Reactor Demonstration Program (ARDP). Elements of the new program include "risk reduction for future development, and advanced reactor safeguards" (DOE-NE 2023). The Office of Nuclear Energy and the National Nuclear Security Administration (NNSA) are working with U.S. industry, under the International Nuclear Security Techniques for Advanced Reactors (INSTAR) program, to foster consideration of advanced security and safeguards concepts early in the design cycle of advanced reactors (National Nuclear Security Administration 2022).

To prevent the theft of nuclear material or sabotage of nuclear facilities, the INSTAR program partners with the U.S. advanced nuclear reactor industry and embarking nuclear power countries on improving the security of future U.S. advanced reactor exports. The focus is on three areas: (1) early integration of security by design; (2) building nuclear security capacity in countries embarking on new nuclear power programs; and (3) strengthening the global nuclear security regime by developing international guidelines and resources on evolving security considerations posed by advanced and small modular reactors. These partnerships help support the responsible international deployment of U.S. advanced reactor technologies while ensuring technological innovation in meeting global security legal obligations and requirements.

The INSTAR program, a congressionally mandated program, provides funding to DOE national laboratory experts to partner with vendors under Cooperative Research and Development Agreement (CRADA) or nondisclosure agreement (NDA). By working with INSTAR, U.S. companies will be better positioned to engage with global customers. Specific areas of support are customized to the vendor's needs, reactor design concept, and technology readiness level.

The DOE/NNSA is mandated by Congress to work closely with the DOE-NE and the U.S. NRC to support the development and integration of security by design in U.S. origin technology. INSTAR seeks to move U.S. civil nuclear technology development forward, recognizing the critical role it plays for meeting climate-change goals. It seeks to accomplish this by integrating security considerations into the vendors' existing processes without hampering development timelines (National Nuclear Security Administration 2022).

Looking ahead, as the number of nuclear reactors increases globally, there will be more opportunities for terrorists to potentially access nuclear materials both at nuclear facilities and while those materials are in transit. Accordingly, there will be more nuclear material, nuclear facilities, and host countries that will need to develop new and advanced approaches to security and safeguards.

Leveraging the U.S. domestic civilian nuclear sector is the most effective means by which high U.S. standards of nuclear safety, security, and safeguards can be exported around the world. The forecasted expansion of this sector in support of U.S. and global clean energy goals, will require broader authorized use of nuclear materials in commerce, which could translate into a potential increase in opportunities for diversion and misuse (National Nuclear Security Administration and U.S. Department of Energy 2021;

U.S. Department of Energy 2023). This makes it all the more imperative to advance a dominant presence of U.S. industry participants in the global nuclear market.

The new nuclear renaissance provides an opportunity for the U.S. government to work with industry and the IAEA to develop appropriate standards to prevent diversion of nuclear material and to keep nuclear terrorism at bay. A coordinated interagency forum, such as a working group or council, could be a mechanism to define issues, identify ongoing related work, coordinate disparate agency efforts, and provide an entry point for existing and new entrant domestic nuclear companies seeking to align with and advance U.S. interests in the sector. The IAEA Small Modular Reactor Regulators' Forum is an example of this type of collaborative dialogue (International Atomic Energy Agency 2023b), where countries seek to achieve additional harmonization on nuclear safety topics between regulatory authorities of the participant countries. Ultimately, the substantive engagement of industry participants will be essential to the development of workable standards for both safety and safeguards, as suggested here.

5.2 EXPANDING GLOBAL NUCLEAR ENERGY

"We can design a system that's proof against accident and stupidity; but we can't design one that's proof against deliberate malice."

Arthur C. Clarke
2001, A Space Odyssey

FINDING 5-1: As a result of increasing reliance on nuclear power, more nuclear material and facilities around the globe will require new methods of security that move beyond relying primarily on guns, guards, and gates to safeguard against nuclear terrorism.

Russian attacks on nuclear power plants and the civil energy sector in Ukraine have broken the long-standing norm that placed nuclear facilities off limits in times of conflict, as set out in the 1949 Geneva Convention, Section A, Rule 42 and Protocol 1, Article 56. Diplomatic efforts through the IAEA and other multinational fora, with support from and led by the United States, should seek to bolster international consensus agreements for reestablishing this norm. Nonetheless, it is unclear whether such agreements would be honored by those committing acts of aggression or terrorism that already violate national sovereignty, treaty commitments, and basic international norms of behavior.

Russia's attacks and the ongoing terrorism risk have generated new concerns about the future targeting of the civil energy infrastructure, particularly nuclear power plants, by state and non-state actors. Such energy facilities, nuclear and non-nuclear, while robust, are not designed or constructed to repel a determined and sustained military assault. A regulatory requirement for existing facilities to withstand military assaults would almost certainly place a prohibitively high economic burden on the sector. However, as new reactor technologies are being developed, there will be opportunities to think about the concept of inherent safety and security that could be included into the design where feasible. Engineers and regulators could explore new options in the design phase and then further analyze options to understand the costs and operational impacts, if any. Promising topical areas already under examination and development include robust nuclear fuel concepts, spent fuel storage technologies, and extended coping times prior to the need for auxiliary power supplies.

5.3 DESIGN BASIS THREAT

FINDING 5-2: Civil nuclear energy facilities are not hardened against a nation-state assault under military conditions of sustained war fighting, and efforts to make them so are likely to be cost-prohibitive. However, additional international coordination could result in more states adopting additional control measures and contingency plans to reduce the resulting risk.

As defined by the U.S. NRC (U.S. Nuclear Regulatory Commission 2023), the design basis threat (DBT) is "a description of the type, composition, and capabilities of an adversary, against which a security system is designed to protect. The NRC uses the DBT as a basis for designing safeguards systems to protect against acts of radiological sabotage and to prevent the theft of special nuclear material." According to the IAEA, a DBT includes the capabilities of potential insider and external threats that include unauthorized removal of nuclear and other radiological material or sabotage of a nuclear facility. Physical protection systems are built to and reviewed on the basis of the DBT (International Atomic Energy Agency 2009, 2023a).

The participation by U.S. government and private sector experts in international, multilateral initiatives, such as the International Atomic Energy Agency Nuclear Harmonization and Standardization Initiative, has made a positive contribution toward achieving the goal of safe and secure deployment of small modular reactors and other advanced nuclear technologies. It also supports maximizing the potential contribution of such technologies for the achievement of global clean energy goals.

The US Department of Energy/National Nuclear Security Administration leads several efforts to strengthen international safeguards. These programs include support to the IAEA for developing technology tools that can advance the more efficient and effective application of safeguards. They also include efforts with partner countries and organizations to reinforce the safeguards regime to include promoting universal adherence to IAEA Additional Protocol. Other efforts include developing approaches for safeguarding new and prospective new fuel cycle facilities, existing facilities, and emerging technologies.

The DOE/NNSA is working to incorporate safeguards-by-design elements into U.S. advanced reactor designs by engaging with nuclear industry stakeholders on improving opportunities for international deployment. The DOE/NNSA also partners with U.S. industry on security by design for advanced and small modular reactors to improve the security of U.S. exports, make U.S. advanced nuclear reactor designs more competitive, and make future markets more resilient to evolving risks (National Nuclear Security Administration 2021a, 2021b). Given the scope and scale of global nuclear ambitions cited above, a significant expansion and acceleration of U.S. civil nuclear development efforts is needed to catch up with the many Russian and Chinese commercial nuclear ventures that are currently underway.

5.4 INTERNATIONAL STANDARDS AND REGULATIONS

FINDING 5-3: The United States is working to drive international standards and regulatory harmonization around nonproliferation, materials control and accounting, and physical- and cybersecurity for advanced nuclear technologies.

The NNSA is the lead technical advisor for U.S. 123 Agreement negotiations and leads implementation of a number of these agreements and implementing arrangements (Arms Control Association 2019; U.S. Department of State 2022). The NNSA also provides robust support for the part 810 process to ensure exports of U.S. technology are not diverted or used for proliferant purposes (National Nuclear Security Administration 2023).

A National Academies study committee recently completed an analysis of the challenges and potential support for the deployment of advanced nuclear reactors. The committee recommended that the United States develop a plan for increased and sustained long-term financial and technical support for capacity-building in partner countries, including cost requirements for using U.S. national laboratories and universities as training platforms. This plan should include partnering with U.S. reactor vendors to develop a safety, safeguards, and security "package." With this package, the United States and the vendor could offer customized support to a host country for developing and implementing new safety, safeguards, and security arrangements (Recommendation 9-5 [National Academies of Sciences 2023b]). Our committee, although not having studied this and related issues in the same depth, endorses this recommendation as supporting efforts to advance nuclear security, at home and abroad.

5.5 THE IMPORTANT ROLE OF THE INTERNATIONAL ATOMIC ENERGY AGENCY

FINDING 5-4: For nuclear new-entrant nations, efforts are urgently needed to bolster nuclear energy cooperation in developing training, education, safety, security, safeguards, and nuclear governance.

FINDING 5-5: The IAEA and other multilateral fora are carrying out effective initiatives for strengthening international standards on nuclear safety, security, and materials control and accounting for advanced nuclear technologies. Unfortunately, the IAEA lacks adequate resources including baseline funding and extra-budgetary contributions to fully address the demands of the expanding civil nuclear sector.

The IAEA small modular reactors (SMR) Regulators' Forum, created in March 2015, provides enabling discussions among member states and other stakeholders to share SMR regulatory knowledge and experience. The forum enhances nuclear safety by identifying and resolving common safety issues that may challenge regulatory reviews associated with SMR and by facilitating robust and thorough regulatory decisions. The forum's work is intended to establish position statements on regulatory issues; suggestions for revisions to or new IAEA documents; information to help regulators enhance regulatory frameworks; reports on regulatory challenges with discussion on paths forward; and suggestions for changes to international codes and standards (International Atomic Energy Agency 2023b, 2023c).

The IAEA has also launched an initiative bringing together policymakers, regulators, designers, vendors, and operators to develop common regulatory and industrial approaches to SMRs. The Nuclear Harmonization and Standardization Initiative (NHSI) aims to facilitate the safe and secure deployment of SMRs and other advanced nuclear technologies to maximize their contribution to achieving decarbonization goals (International Atomic Energy Agency 2023c).

The increasing presence of nuclear security in IAEA activities has spread to encompass the IAEA's work on new and advanced reactors. In 2021, the IAEA began a nuclear security project to share information on SMR security systems and how requirements and guidance from the Nuclear Security Series can apply to SMRs (International Atomic Energy Agency 2021). This project will form the basis of future guidance and training programs.

The current regular budget for IAEA nuclear security and safeguard activities is €6.4 million for nuclear security and €133.5 million for safeguards implementation (International Atomic Energy Agency 2019). The IAEA Division of Nuclear Security relies on extra-budgetary funding five or six times the amount of the regular budget to conduct its work (U.S. Government Accountability Office 2019). Furthermore, given the zero-growth constraints on the IAEA budget (International Atomic Energy Agency 2022; U.S. Government Accountability Office 2019, p. 10) and an unwillingness by some IAEA member states to reapportion the budget from other activities (see U.S. Government Accountability Office 2019, pp. 30–32), efforts to make deployment of new and advanced reactors safeguarded and secure will be impaired unless there are significant changes in the current funding stream. These initiatives that aid potential partners for deploying U.S. new and advanced reactors, either bilaterally through U.S. government initiatives or multilaterally through the work of the IAEA, will require stepped-up efforts by the United States and the IAEA. The IAEA will need a considerable increase in its budget to meet the safety, security, and safeguards objectives for these new and expanded nuclear programs (National Academies of Sciences 2023a, 2023b).

5.6 SPENT NUCLEAR FUEL STORAGE

FINDING 5-6: The challenge of mitigating the domestic security risk of nuclear materials would be advanced by addressing the long-term disposal issue of spent nuclear fuel that is currently stored at nuclear power plants and other facilities across the United States. Addressing the need to permanently store spent fuel would reduce the risk of these nuclear materials being obtained or targeted by terrorists.

Faced with a burgeoning nuclear power program in the 1970s, the attention of U.S. policymakers eventually turned to resolving the long-term disposal issue of spent nuclear fuel being stored at nuclear power plants. This work culminated in the passage of the Nuclear Waste Policy Act of 1982 as amended, which designated the Yucca Mountain site in Nevada as the first disposal location. The Act also created a fee to be collected from electricity consumers to fund the development of the disposal program. Although over $44 billion has been collected from consumers, four decades after passage of this legislation, the United States is no closer to a functioning program for spent fuel disposal.

Many studies have been completed on this issue by committees, commissions, and panels. This committee did not recreate that work but notes the obvious. The absence of a long-term disposal solution leaves spent nuclear fuel housed at multiple locations throughout the United States, including locations at long-dismantled former nuclear power installations that, but for the absence of a disposal or consolidated storage site for their spent fuel, could long since have been turned into brown field locations and returned to other uses. From a security perspective, this situation is broadly undesirable because it results in many more locations that must be secured and monitored. This committee endorses the result of many earlier expert panels that have recommended that the long-standing political impasse be resolved for the benefit of the nation and its security. The committee also notes that in December 2021 the DOE issued a request for information seeking feedback on consent-based siting as an approach to managing interim storage of spent nuclear fuel.

RECOMMENDATION 5-1: A whole-of-government effort, in partnership with the civil nuclear sector, is needed to strengthen the U.S. presence in civil nuclear energy commerce and thereby enhance global standards for safety, security, and materials control.

This effort could include:

- Establishing an inter-agency working group to coordinate ongoing U.S. government activities related to civil nuclear exports strategy and to promote regulatory harmonization of safety, safeguards, and security standards and licensing frameworks by:
  1. Seeking additional opportunities to facilitate international nuclear energy cooperation for developing training, education, safety, security, safeguards, and nuclear governance required for nuclear new-entrant nations, while defining U.S. priorities and eliminating duplication of effort among agencies; and
  2. Identifying gaps in international efforts and making recommendations for U.S. extra-budgetary contributions to the IAEA or other multilateral fora, and advocate for similar actions by allies and partners. The objective would

be to advance and accelerate the development of global standards and other harmonization and updates to national and international frameworks.

- Encouraging the U.S. Nuclear Regulatory Commission to move beyond its traditional reticence in leading international regulatory harmonization efforts. This effort should be animated by a recognition that such efforts do not necessarily constitute technology advocacy, need not impair individual national sovereignty, would promote the attainment of higher global standards of safety and security, and are crucial to U.S. national security in an increasingly nuclearized world.

References

Ahn, Alan, and Todd Allen. 2022. "2022 Advanced Nuclear Map: Charting a Breakout Year." http://thirdway.imgix.net/pdfs/2022-advanced-nuclear-map-charting-a-breakout-year.pdf.

Arms Control Association. 2019. "The U.S. Atomic Energy Act Section 123 at a Glance." https://www.armscontrol.org/factsheets/AEASection123.

Holt, Katherine. 2021. "Advanced Reactor Civil Nuclear Security Project." Conference: Proposed for presentation at the National Academies of Science Meeting—Merits and Viability of Different Nuclear Fuel Cycles and Technology Options and the Waste Aspects of Advance held May 18, 2021, in Washington, DC. https://www.osti.gov/biblio/1868438.

International Atomic Energy Agency. 2009. *Development, Use and Maintenance of the Design Basis Threat. IAEA Nuclear Security Series No. 10.*

---. 2019. *The Agency's Programme and Budget 2020–2021.*

---. 2021. *Nuclear Security Report 2021.*

---. 2022. *The Agency's Budget Update for 2023.*

---. 2023a. "Design Basis Threat (DBT)." https://www.iaea.org/topics/security-of-nuclear-and-other-radioactive-material/design-basis-threat.

---. 2023b. "Small Modular Reactor (SMR) Regulators' Forum." https://www.iaea.org/topics/small-modular-reactors/smr-regulators-forum.

---. 2023c. "The SMR Platform and Nuclear Harmonization and Standardization Initiative (NHSI)." https://www.iaea.org/services/key-programmes/smr-platforms-nhsi.

International Energy Agency. 2022. *World Energy Outlook 2022.* https://www.iea.org/reports/world-energy-outlook-2022

National Academies of Sciences, Engineering, Medicine. 2023a. *Laying the Foundation for New and Advanced Nuclear Reactors in the United States.* Washington, DC: The National Academies Press.

---. 2023b. *Merits and Viability of Different Nuclear Fuel Cycles and Technology Options and the Waste Aspects of Advanced Nuclear Reactors.* Washington, DC: The National Academies Press.

National Nuclear Security Administration. 2021a. "NNSA: Reducing Global Nuclear Threats." https://www.energy.gov/sites/default/files/2021-12/A%29%20NNSA_ReducingGlobalNuclearThreats.pdf

---. 2021b. "Prevent, Counter, and Respond—A Strategic Plan to Reduce Global Nuclear Threats (NPCR)." https://www.energy.gov/nnsa/articles/prevent-counter-and-respond-strategic-plan-reduce-global-nuclear-threats-npcr.

---. 2022. *International Nuclear Security Techniques for Advanced Reactors (INSTAR).* Edited by International Nuclear Security.

---. 2023. "10 CFR Part 810." https://www.energy.gov/nnsa/10-cfr-part-810.

National Nuclear Security Administration and U.S. Department of Energy. 2021. *Prevent, Counter, and Respond—NNSA's Plan to Reduce Global Nuclear Threats*. FY 2022-FY 2026.

Nuclear Energy Institute. 2022. "Advanced Nuclear Energy." https://www.nei.org/fundamentals.

U.S. Department of Energy. March 2023. *Department of Energy FY 2023 Budget in Brief*. https://www.energy.gov/sites/default/files/2023-06/doe-fy2024-budget-in-brief-v5.pdf.

U.S. Department of Energy Office of Nuclear Energy (DOE-NE). 2023. "Nuclear Reactor Technologies." https://www.energy.gov/ne/nuclear-reactor-technologies.

U.S. Department of State. 2022. "123 Agreements." https://www.state.gov/fact-sheets-bureau-of-international-security-and-nonproliferation/123-agreements/.

U.S. Government Accountability Office. 2019. *Nuclear Security: The International Atomic Energy Agency Could Improve Priority Setting, Performance Measures, and Funding Stabilization*.

U.S. Nuclear Regulatory Commission. 2023. "Design-Basis Threat (DBT)." https://www.nrc.gov/reading-rm/basic-ref/glossary/design-basis-threat-dbt.html.

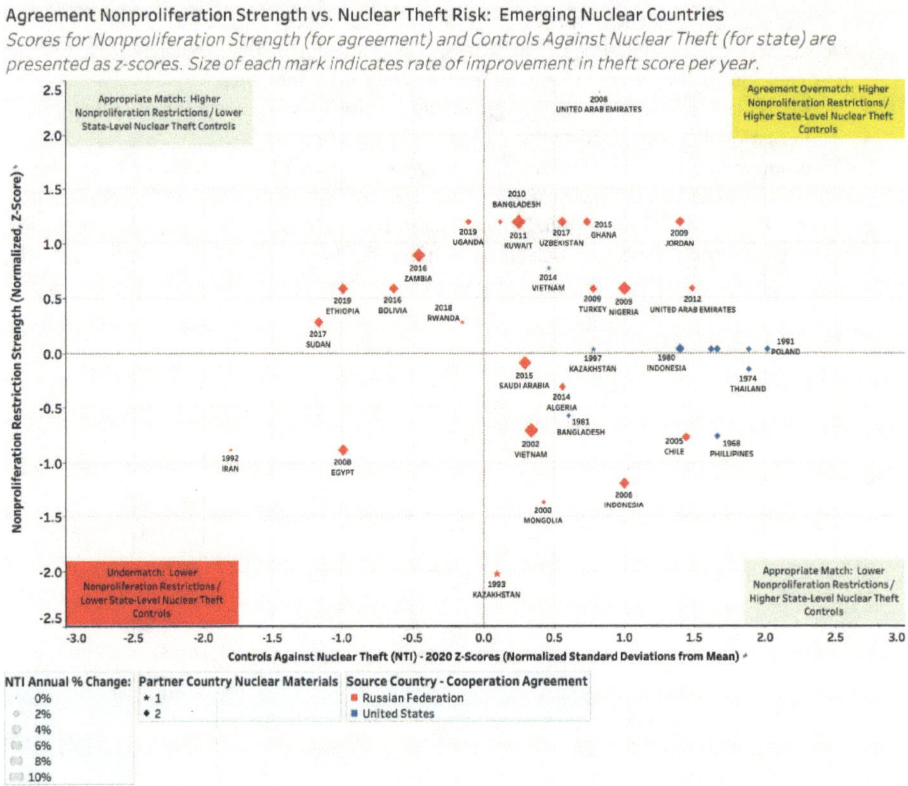

FIGURE 6-1 A plot of the strength of nonproliferation restrictions (y-axis) and state controls against the theft of nuclear materials (x-axis). Researchers coupled data on nonproliferation restriction strength with the Nuclear Security Index developed by the Nuclear Threat Initiative (Bidgood and Potter 2021). The results show that the agreements and nuclear export controls are best for nuclear newcomers in the upper right quadrant, with worrisome agreements and materials control for countries in the lower left. The lower right quadrant depicts countries identified with above-average materials control and below-average agreements, with the opposite true in the upper left quadrant, lower-than-average materials control and above-average agreements. A qualitative measure of the amount of nuclear material possessed by a country is represented by the size of the diamond data points. The differences between cooperative agreements with Russia or the United States are a reflection of the changing political dynamic between the two countries discussed in chapter 4.

SOURCE: Dr. William C. Potter, Middlebury Institute of International Studies at Monterey.

6

The Risks Associated with Highly Enriched Uranium and Plutonium

Consolidating nuclear weapons and nuclear weapons-usable material to fewer sites is one of the most effective strategies for reducing the risk of nuclear theft. Every facility that eliminates its nuclear weapons, highly-enriched uranium (HEU), or separated plutonium is one less potential target that needs to be protected from theft or sabotage (Bunn, Roth, and Tobey 2019).

Global stocks of civilian and military weapons-usable material are large. As of May 2022, there were approximately 1,250 tons of HEU (+/- ~125) and 550 (+/- ~10) tons of separated plutonium in more than 20 countries. While quantities of HEU are slowly declining, there has been a growth in plutonium stockpiles (International Panel on Fissile Materials 2022a, 2022b). This increase is associated with the reprocessing of spent nuclear fuel by the civilian nuclear energy sector.

There are long-standing gaps in efforts to consolidate weapons-usable nuclear materials. With the notable exception of the Cooperative Threat Reduction efforts with the former Soviet Union, the United States has largely focused on consolidating and minimizing civilian materials. Currently, there is no dedicated program focused on consolidation and minimization of military materials internationally. Additionally, while the United States has supported removal of plutonium from a small number of research facilities, there has been much less focus on consolidating or limiting the growth of larger civilian plutonium stocks.

Of note, the United States has also supported stronger norms on minimization of weapons-usable nuclear material at the multilateral level. In 2016, the United States joined nearly two dozen countries in committing to concrete steps to consolidate HEU, curtail its use in the civil sector, and report inventories (Nuclear Security Summit 2016). The commitment has been formalized and opened to all states as International Atomic Energy Agency (IAEA) Information Circular 912 (INFCIRC/912). Unfortunately, only two countries have followed through on their INFCIRC/912 commitments by releasing

BOX 6-1
Summary

Since the end of the Nuclear Security Summit process in 2016, there has been a gradual slowing of efforts to eliminate excess civilian stockpiles of highly enriched uranium (HEU) and separated plutonium. While the inventories of HEU have declined slightly, since 2020, the inventories of separated civil plutonium have increased by more than 17,000 kilograms. Non-state actors, both domestic and international, currently lack the technical ability to create these materials. However, given the interest by some in obtaining them, it should be a top national security priority to eliminate these materials wherever possible, and better secure those materials that remain by the 22 countries that possess them.

Highlights

- Efforts to eliminate, reduce, and secure the most dangerous civilian special nuclear materials must be prioritized as a national security objective.
- Stored amounts of highly enriched uranium have been reduced over the past 15 years, and efforts should be redoubled to further reduce excess HEU globally.
- Separated plutonium remains a concern as inventories are increasing. Safe and secure storage of nuclear materials (including HEU) at the highest standard remains challenging.
- Plutonium should not be used as a fuel source in the expanding civilian nuclear energy sector.

information about their civilian HEU quantities (International Atomic Energy Agency 2020).

6.1 SPECIAL NUCLEAR MATERIAL

Special nuclear materials, specifically HEU and separated plutonium, are the primary ingredients for a nuclear weapon. Because these are manmade, significant industrial facilities and scientific expertise are required to manufacture these materials to produce them in high quantities.[1] Constructing and operating these facilities are outside the reach of non-state actors. If a non-state actor was able to gain access to already manufactured HEU, however, the scientific expertise to make a crude improvised nuclear device (IND) may be within the reach of a group intent on carrying out

[1] Fortunately, this material largely does not exist in nature, but the committee notes there is one well-known source of plutonium reported in the research literature (Krieger 2004).

a terrorist attack (Bunn and Wier 2006). Any such event would have profound global repercussions, ranging from the direct damages to the location where the detonation occurred to cascading economic and psychological consequences as nations struggle to reassure their publics throughout the response and recovery from such an incident. With today's pervasive social media, any response and recovery efforts would no doubt be carried out in an environment of considerable misinformation and disinformation. For the civil nuclear sector, an IND attack may reverse or even halt the expected expansion of nuclear energy to help combat global warming. As such, the focus for government efforts has been and should continue to be on the elimination of as much special nuclear material as possible. As noted earlier in this report, multiple terrorist organizations have sought weapons of mass destruction, including nuclear and radiological weapons. With an increase in the range of ideologies interested in creating mass chaos and with some religious groups committed to apocalyptic goals, protecting and securing weapons-quality materials must be an urgent priority for governments around the world. Every country has a role to play.

6.1.1 Highly Enriched Uranium

FINDING 6-1: Acquiring sufficient quantities of HEU is the most difficult step in constructing an IND. Protecting and minimizing this material is the cornerstone of the NNSA mission in preventing nuclear terrorism. Further reductions could be realized in partnerships with those countries with excess HEU.

Prior to 2020, there was a significant reduction in the use of HEU in the civilian sector (Table 6.1). With the removal of all HEU from Nigeria in 2018, 32 countries plus Taiwan have removed all such material from their territories (National Nuclear Security Administration 2018). Overall, more than 7,000 kilograms of special nuclear materials have been removed from over 100 facilities. However, more work remains to be done as there are still 22 countries that possess at least 1 kilogram of HEU or separated plutonium. The IAEA notes that as little as 25 kilograms of HEU would be sufficient to create a nuclear device.

The United States has been the most active country in encouraging other countries to eliminate their stocks of special nuclear material. Initial DOE/NNSA programs included the Materials Consolidation and Conversion program. The aim on this effort was to consolidate and down-blend Russian highly enriched uranium. Additionally, a dedicated U.S. program to convert the remaining research reactors, domestic and international, from HEU fuel to low enriched uranium (LEU) fuels has made impressive scientific progress in the development of new LEU fuels. A 20+ year effort with Russia to repatriate Russian-origin HEU from third countries was very successful.[2] Another U.S. program is focused on removing excess HEU from reactors in countries across the globe and repatriating it to the country where that material originated. This work

[2] Russia is no longer participating in this program.

TABLE 6-1. Countries That Have Eliminated Weapons-Usable Nuclear Material

Year	Country or Territory	(cont.)	
1992	Iraq	2009	Libya, Romania, Taiwan
1996	Columbia	2010	Chile, Serbia
1997	Spain	2012	Mexico, Ukraine, Sweden, Austria
1998	Denmark	2013	The Czech Republic, Vietnam
1999	Thailand, Slovenia, Brazil, Philippines	2015	Hungary, Jamaica, Uzbekistan, Georgia
2005	Greece	2016	Argentina, Indonesia, Poland
2007	South Korea	2017	Ghana
2008	Latvia, Bulgaria	2018	Nigeria

SOURCE: Table 1, p. 70, in Bunn, Matthew; Roth, Nickolas; and Tobey, William H. *Revitalizing Nuclear Security in an Era of Uncertainty*. Report. Harvard Kennedy School of Government. Belfer Center for Science and International Affairs. January, 2019.

was accelerated during the Obama Administration's Nuclear Security Summits, which provided important high-level impetus to complete this work.

Norway has also provided important leadership through its efforts to strengthen international norms on the non-use of HEU in civilian applications. It has hosted three HEU Minimization Symposiums, the most recent taking place in 2018 (Arms Control Association 2018). Further, Norway sponsored an IAEA Information Circular in 2017 where countries commit to reduce their stocks of HEU and provide regular reporting on the current inventory of material. Norway is also taking steps to dispose of its modest stockpile of HEU in the coming years.

More work can be done to reduce and eliminate HEU from civilian applications globally. Approximately 85 facilities are still operating using HEU, with 56 located in Russia alone. There are several countries, including South Africa and Belarus, where there are hundreds of kilograms of excess HEU that could be eliminated. There are serious political, economic, and technical challenges that need to be confronted, but with strong U.S. leadership generating the necessary political will, these can be overcome by the global community.

6.1.2 Plutonium

FINDING 6-2: Inventories of separated plutonium are increasing worldwide, elevating the amount of materials that could potentially find their way to proliferant states or non-state actors.

In stark contrast to the work done to reduce HEU, there has not yet been a concerted effort to reduce civilian stockpiles of separated plutonium. Over the past 10 years, based on reporting to the IAEA, the world has seen an increase in separated plutonium of 51 metric tons, bringing the total inventories to a staggering 371 metric tons. Using the IAEA assessment that only 8 kilograms of separated plutonium are required to make a nuclear device, that is enough to make more than 46,000 nuclear devices.

There is a different dynamic at play with stockpiles of separated plutonium. Instead of it being used for scientific experiments, certain countries have chosen to use plutonium in nuclear reactor fuel. In particular, five of the 31 countries with active nuclear programs have gone this route: China, France, India, Japan, and Russia. By using this type of fuel cycle, a country must reprocess spent fuel to acquire the plutonium for the fuel (or contract with another country to do so). This is the same process that a country would pursue to acquire plutonium for nuclear weapons. Therefore, the fewer countries that reprocess spent fuel, the fewer countries there will be to have the capacity to create nuclear material for a nuclear weapon. Given that 26 countries have demonstrated that they can forego using separated plutonium as a source of nuclear energy, it should be possible for other countries do so as well.

The technical capabilities exist to eliminate special nuclear materials in several countries around the world. In the case of unirradiated HEU, this material can be down-blended into LEU and then used as fuel for nuclear energy.

Irradiated HEU and plutonium are more difficult to dispose of, given the high levels of radiation. Efforts to identify long-term solutions for dispositioning these materials, as well as spent fuel in general, have floundered in the case of Yucca Mountain in the United States. A few countries in Europe—including Finland and Sweden—have made recent progress on commissioning a long-term repository for such materials (i.e., spent fuel), but more progress is needed, especially given the expected expansion of civil nuclear energy to help combat climate change.

6.2 SECURITY OF NUCLEAR MATERIALS

In locations where civilian stocks of special nuclear materials remain, implementing robust nuclear security measures must be a priority. This includes advancing a variety of measures that help to protect the facilities where the nuclear materials are stored, account for the inventories of nuclear materials to support detection of possible diversion, and respond to insider attempts to steal these materials or forcible entry attempts to obtain them. Underpinning these measures must be a strong nuclear security culture where employees remain vigilant to the threats posed to their facilities.

The contemporary challenges for securing excess nuclear materials are much different from those faced in the immediate aftermath of the fall of the Soviet Union. Some have argued for a new international framework to guide how to manage these materials in the long run (Nunn and Holgate 2021). But global nuclear security can only be strengthened if countries work on a peer-to-peer level to secure and manage these materials they are not able to eliminate. Equally important is developing safeguards and security measures that take into account technology advances, such as artificial intelligence, and how these technologies can both help protect and potentially be exploited by adversaries to access these materials.

6.2.1 2023 Nuclear Security Index

FINDING 6-3: Excess nuclear materials have not been secured to the highest standard and consolidated to as few locations as possible.

Despite the United States' role in strengthening nuclear security around the world, significant vulnerabilities remain at nuclear sites and within the international institutions and legal frameworks that support nuclear security. The Nuclear Threat Initiative's 2023 Nuclear Security Index, which is the most comprehensive public quantitative assessment of nuclear security conditions in 175 countries and Taiwan, found that, since 2020, nuclear security has been regressing in "countries and areas with the greatest responsibility for preventing nuclear theft and sabotage—those with nuclear materials and facilities" (Nuclear Threat Initiative 2023). The study warns about the dangers of rapidly growing stocks of separated plutonium as a result of commercial reprocessing; insufficient progress strengthening security culture and insider threat prevention programs in countries with weapons-usable nuclear materials and nuclear facilities; and faltering support for new commitments and assurances that increase international confidence in the effectiveness of nuclear security.

The report also highlights long-standing gaps in nuclear security around the world. Among the 46 countries and Taiwan with nuclear facilities, 25 have no regulations or licensing conditions that require personnel to report suspicious behavior; 31 do not require drug testing, background checks, and psychological and mental fitness checks for personnel; and 17 do not have regulations that require the use of a regularly updated design basis threat. Particularly alarming in this new era of evolving risks is the report finding that 16 of the 46 countries and Taiwan with nuclear facilities do not require plans for protecting nuclear infrastructure during a natural or human-caused disaster.

The United States has now entered a new era where dangerous new risks are intersecting with long-standing vulnerabilities. For decades, U.S. leadership has led to stronger nuclear security internationally. This leadership is now more important than ever as countries respond to rapidly evolving nuclear security threats and challenges to nuclear security operations.

RECOMMENDATION 6-1: The United States should prioritize the effort to secure, and wherever practical, consolidate or eliminate civilian special nuclear materials and treat it as a core national security objective. This includes leading efforts to transform perspectives on the use of plutonium for nuclear energy production.

The United States has a long-standing record of providing the essential leadership for addressing the risks associated with nuclear terrorism. However, in the years since the Nuclear Security Summits ended in 2016, the United States has been increasingly prioritizing regional considerations, which has, at times, come at the expense of following through on key international nuclear security activities. President Biden's release of National Security Memorandum 19 (NSM 19), which outlines the Administration's strategy to counter weapons of mass destruction terrorism, expresses a commitment to reprioritize these global activities as vital to U.S. national security (The White House 2023).

Beyond the important actions outlined in NSM 19, there are additional steps the U.S. government can take to reassert its leadership role. The United States has unique facilities that are able to dispose of special nuclear materials. These facilities should be fully harnessed to repatriate these materials for disposal before they reach the end of their useful life. This will require a focused effort to strengthen international cooperation on addressing this issue.

There is important domestic work to be done as well. There are many metric tons of excess civilian HEU and separated plutonium in the United States that should be eliminated as soon as possible. In instances where these materials are needed, security measures should be evaluated against the existing threat environment for any needed enhancements. Given the growing domestic terrorism challenge and continued evidence of non-state actor interest in weapons of mass destruction, this should include updating insider threat mitigation programs.

Looking to the future, the United States should not start new civilian facilities or activities that utilize HEU, if there are alternative technologies that can be used to achieve the same goals. This is key to safeguarding U.S. credibility in leading international efforts to reduce the amount of HEU material other countries are using. Further, a nonproliferation impact assessment should be required whenever the use of HEU or plutonium is being considered. Similarly, for the potential use of HEU in space applications, there should be assessments that consider nonproliferation concerns in the development of those assets. Reducing the quantities of these materials will also result in long-term cost savings on security measures required to protect them (Bunn and Harrell 2012).

The United States has had a well-established commitment to HEU minimization in support of nonproliferation. Recently, DOE has provided funding to start a new experimental HEU-fueled facility at Idaho National Laboratory (INL) known as the Molten Chloride Reactor Experiment (Kramer 2023). While this experiment will utilize existing stockpiles of HEU from a prior research reactor that ran at INL from 1969 to 1990 and will allow the DOE to collect the desired data at reduced cost and scale, it will, nevertheless, mark the first time in more than ten years where a country has begun a civilian activity utilizing HEU (Idaho National Laboratory 2023; Office of NEPA Policy and Compliance 2023). There was no official review of the nonproliferation impact of this project, even though it has implications for U.S. leadership on nonproliferation efforts abroad.

6.2.3 Proliferation Concerns and Nuclear Fuel Cycles

Given the increasing quantities of separated plutonium, a new approach is needed to reduce the risk associated with fissile material. In countries with nuclear energy programs that utilize a plutonium fuel cycle, the conversation focuses on how best to manage these materials. The right approach to pursue is advancing efforts to reduce the attractiveness of these materials by reducing the stockpiles and physically protecting and safeguarding materials so that they cannot be weaponized. From there, countries can prioritize better balancing the supply and demand of plutonium so that it is reduced over time.

Although this committee did not undertake an independent review and validation of proliferation concerns with fuel cycles and developments in nuclear energy, similar and parallel concerns were examined in the National Academies' study "Merits and Viability of Different Nuclear Fuel Cycles and Technology Options and the Waste Aspects of Advanced Nuclear Reactors" (National Academies of Sciences 2023). This report concluded that all advanced reactor fuel cycles will necessitate stringent safeguards and security measures that align with the potential risks they present. There are particular issues with fuel cycles involving the reprocessing and separation of fissile materials. These pose more significant risks in terms of proliferation and terrorism compared to the once-through uranium fuel cycle, where spent fuel is directly disposed. The concern is that separated fissile materials from reprocessing would not be uniformly mixed with highly radioactive fission products and could potentially be used in weapons. The materials that could be separated and potentially used in weapons might include fissionable materials other than the conventional special nuclear materials, HEU and plutonium. Consequently, closed fuel cycles of this nature will likely necessitate specific safeguard technologies to fulfill the IAEA's objective of prompt detection and monitoring.

The National Academies report stated that "the U.S. government should support the IAEA's development and application of effective safeguards for advanced reactor technologies by authorizing, via the U.S. interagency process, IAEA access through the eligible facilities list, especially to those advanced reactor systems for which the IAEA does not currently have safeguards experience. Developers of these types of advanced reactors and fuel cycle facilities should provide facility information to the IAEA to help with integration of safeguards considerations into the design process." (National Academies of Sciences 2023, pp. 13 and 183) Additionally, the NRC ought to address security and material accounting measures for high-assay low-enriched uranium and other attractive nuclear materials that may be present in advanced reactor fuel cycles.

Action on the findings and recommendations from the study "Merits and Viability of Different Nuclear Fuel Cycles and Technology Options and the Waste Aspects of Advanced Nuclear Reactors" on nonproliferation and security risks with nuclear material would address and ameliorate concerns of nuclear terrorism related to separated materials.

To summarize, countries exploring the possibility of utilizing nuclear energy to combat climate change need to avoid options that include separation or use of plutonium and put in place plans for long-term management of spent nuclear fuel.

References

Arms Control Association. 2018. "Remarks: Reducing Stocks of Highly Enriched Uranium." *Arms Control Today*. https://www.armscontrol.org/act/2018-07/features/remarks-reducing-stocks-highly-enriched-uranium.

Bidgood, Sarah, and William C. Potter. 2021. *End of an Era. The United States, Russia, and Nuclear Nonproliferation*. James Martin Center for Nonproliferation Studies, Middlebury Institute of International Studies.

Bunn, Matthew, and Eben Harrell. 2012. "Consolidation: Thwarting Nuclear Theft." *Project on Managing the Atom*. https://www.belfercenter.org/sites/default/files/legacy/files/Consolidation_Thwarting_Nuclear_Theft_corrected.pdf.

Bunn, Matthew, Nickolas Roth, and William H. Tobey. 2019. "Revitalizing Nuclear Security in an Era of Uncertainty." *Project on Managing the Atom*. https://www.belfercenter.org/publication/revitalizing-nuclear-security-era-uncertainty.

Bunn, Matthew, and Anthony Wier. 2006. "Terrorist Nuclear Weapon Construction: How Difficult?" *The Annals of the American Academy of Political and Social Science* 607: 133–149. http://www.jstor.org/stable/25097844.

Idaho National Laboratory. 2023. *Molten Salt Reactors*. https://inl.gov/document/molten-salt-reactors/ Accessed August 17, 2023.

International Atomic Energy Agency. 2020. "Communication Dated 23 January 2020 Received from the Permanent Mission of Australia Concerning the Joint Statement on Minimising and Eliminating the Use of Highly Enriched Uranium in Civilian Applications." INFCIRC/912/Add.4. https://www.iaea.org/sites/default/files/publications/documents/infcircs/2017/infcirc912a4.pdf.

International Panel on Fissile Materials. 2022a. "Materials: Highly Enriched Uranium." https://fissilematerials.org/materials/heu.html.

---. 2022b. "Materials: Plutonium." https://fissilematerials.org/materials/plutonium.html.

Kramer, D. 2023. "Idaho Reactor Project Plans to Use Bomb-Grade Uranium" FYI: Science Policy News. https://ww2.aip.org/fyi/2023/idaho-reactor-project-plans-use-bomb-grade-uranium

Krieger, Kim. 2004. "Natural Nuclear Reactor Explained." *Science*. https://doi.org/10.1126/article.34941. https://www.science.org/content/article/natural-nuclear-reactor-explained.

National Academies of Sciences, Engineering, Medicine. 2023. *Merits and Viability of Different Nuclear Fuel Cycles and Technology Options and the Waste Aspects of Advanced Nuclear Reactors*. Washington, DC: The National Academies Press.

National Nuclear Security Administration. 2018. NNSA Removes All Highly Enriched Uranium from Nigeria.

Nuclear Security Summit. 2016. "NSS 2016: Gift Basket on Minimizing and Eliminating the Use of Highly Enriched Uranium in Civilian Applications." *Joint Statement on HEU Minimization*. http://www.nss2016.org/document-center-docs/2016/4/1/joint-statement-on-heu-minimization-gb.

Nuclear Threat Initiative. 2023. *NTI Nuclear Security Index: Falling Short in a Dangerous World*. In *NTI Nuclear Security Index* https://www.ntiindex.org/wp-content/uploads/2023/07/2023_NTI-Index_Report.pdf.

Nunn, Sam, and Laura S. H. Holgate. 2021. "Cooperative Risk Management and Reduction: A New Framework for Nuclear Materials Security." https://www.nti.org/analysis/articles/cooperative-risk-management-and-reduction-a-new-framework-for-nuclear-materials-security/.

Office of NEPA Policy and Compliance. 2023. "DOE/EA-2209: Molten Chloride Reactor Experiment (MCRE) Project; Idaho Falls, ID." Department of Energy. https://www.energy.gov/nepa/doeea-2209-molten-chloride-reactor-experiment-mcre-project-idaho-falls-id.

The White House. 2023. "Fact Sheet: President Biden Signs National Security Memorandum to Counter Weapons of Mass Destruction Terrorism and Advance Nuclear and Radioactive Material Security. https://www.whitehouse.gov/briefing-room/statements-releases/2023/03/02/fact-sheet-president-biden-signs-national-security-memorandum-to-counter-weapons-of-mass-destruction-terrorism-and-advance-nuclear-and-radioactive-material-security/

FIGURE 7.1 Workers remove a cesium-137 irradiator from Medstar Georgetown University Hospital in 2018. Efforts such as this have been highly effective in decreasing the number of at-risk radiological sources in applications where safer alternatives are available.
SOURCE: Image NNSA, https://www.aip.org/fyi/2021/final-fy21-appropriations-national-nuclear-security-administration.

7

Managing the Risks and Benefits of Radioactive Sources

"We'll also work together to lock down fissile and radiological material to prevent terrorist groups from acquiring or using them."

President Joseph Biden
February 19, 2021

7.1 RADIOACTIVE SOURCES—RISKS AND BENEFITS

FINDING 7-1: Radioactive sources provide ionizing radiation for many beneficial services such as cancer treatment, blood irradiation, sterilization, oil prospecting, medical research, calibration of dosimeters, food safety, and radiography. However, the radiological materials in these sources pose a nuclear terrorism risk since these sources can be stolen. For example, americium-241, cobalt-60, cesium-137, or iridium-192 can be extracted and then used in an radiological dispersal device (RDD) or radiological exposure device (RED).

Radioactive sources provide ionizing radiation for many beneficial services such as cancer treatment, blood irradiation, sterilization, oil prospecting, medical research, calibration of dosimeters, food safety, and radiography. These sources are manufactured products that contain radionuclides that emit gamma rays, alpha particles, beta particles, or neutrons. The radionuclides are typically encapsulated inside the source to prevent accidental release of the radiological material. Encapsulation can be in the form of layered stainless steel. Common radionuclides used in sources showing the half-lives and type of radiation emitted include: americium-241 (432.2 years; principally alpha radiation with weak gamma ray; neutron radiation when combined with beryllium), cesium-137 (30.17 years; strong gamma ray with beta radiation), cobalt-60 (5.27 years; two strong gamma rays), iridium-192 (74 days; beta and gamma radiation), and strontium-90 (29 years; beta radiation).

BOX 7-1
Summary

Radioactive sources found in commonly used tools and equipment can be used in a radiological dispersal device (RDD) or a radiological exposure device (RED). As such, more attention should be directed to mobilizing and sustaining efforts to identifying technological alternatives to these materials, raising awareness of the risk, and enacting stronger measures for safeguards and security. This should include working with industry and international partners to close gaps in detecting illicit trafficking along the various pathways that terrorist groups might exploit.

Highlights

- Radioactive sources provide ionizing radiation for many beneficial services, such as cancer treatment, blood irradiation, and sterilization, but pose a nuclear terrorism risk since these sources can be stolen and used in an RDD or RED.
- The National Nuclear Security Administration (NNSA) has effective programs in place to support the development and deployment of alternative technologies to replace radioactive sources, taking into account the need for cost-effective devices for providing beneficial services.
- Emerging technology, such as artificial intelligence and machine learning, have risks and benefits to the U.S.-led efforts to address the terrorism risk posed by RDD and RED.

There are many common but shorter-lived isotopes used in medical diagnostics. Most radionuclides have short half-lives (typically less than several days) or long half-lives (more than several hundred years). Short-half-life materials decay rapidly and do not last long enough to pose a radiological contamination threat. Long-half-life materials emit their radiation slowly and thus pose a lower radiation hazard.

Of the common radionuclides used in sources, two are among the most frequently used and both present a risk should terrorists use them in an RDD or RED and are a significant security concern: cesium-137 and cobalt-60. According to the 2021 National Academies study on radioactive sources, the United States has approximately 72,000 of the higher-activity Category 1 and Category 2 cobalt-60 sources, accounting for about 90 percent of all Category 1 and 2 sources in the United States; there are approximately 3,200 cesium-137 sources in these categories, accounting for about 4 percent of all such sources in the United States (National Academies of Sciences 2021) (the categories are defined below).

Cobalt-60, a solid metal, is used in radiation therapy either in implants or as an external source for exposure (Jefferson Lab Resources 2023). It is also used in industry in irradiating medical instruments and in food sterilization. In addition, cobalt-60's radiation is used in the sterile insect technique for control of pests that can infect crops. Cobalt-60 teletherapy units are being phased out in high-income countries as new technologies are adopted, but they remain in wide use in many low- to middle-income countries (LMIC) (Oncology Medical Physics 2023). The 2021 National Academies study identified a number of barriers to replacing cobalt-60 teletherapy in LMIC with non-radioisotopic linear accelerators (LINACs). In particular, lack of reliable electricity and access to technical resources to maintain LINACs can hinder adoption of this alternative technology.

Cesium-137 is usually in the chemical form of cesium chloride, a powder-like "salt" substance that is easily dispersible if removed from its encapsulation. Cesium chloride (CsCl) is commonly used in blood irradiation and in research irradiators. Because these devices can be found in so many domestic and international locations (e.g., blood blanks), ensuring that they are adequately safeguarded is challenging. A prominent example of both the danger and security challenge that CsCl represents is highlighted in an incident that took place in Goiania, Brazil, in September 1987 (International Atomic Energy Agency 1988). Two individuals who were looking for scrap metal to sell broke into an abandoned radiotherapy institute and stole a teletherapy unit, not knowing there was a radiological risk. The unit was accidentally damaged, breaking the encapsulated sealed source containing the CsCl, and as a result of the dispersal, four people died from radiation sickness and more than 249 were contaminated, internally or externally. The impact resulted in more than 112,000 people needing to be monitored and tens of millions of dollars spent over three years to complete the cleanup (International Atomic Energy Agency 1988).

As noted above, CsCl is a salt-like substance, which means it can dissolve and penetrate soil and other geological materials. The cesium can also bind to concrete in buildings, raising the difficulty and cost of cleanup. This is why the risk of an RDD containing high-explosive (HE) materials combined with the CsCl, perhaps stolen from a local blood bank, is so concerning. Should such a device be detonated in an urban environment, contaminating the buildings and critical infrastructure in the vicinity, the disruption to the community and associated economic costs could be enormous.

In addition to half-life and the types of emitted radiation, the amount of radioactivity in a source is an important consideration for developing and implementing safety and security protocols. The International Atomic Energy Agency (IAEA) has developed and published a categorization scheme with five categories, as shown in Box 7-2 (International Atomic Energy Agency 2005). Each category is defined by the potential harm to the health of people should they be exposed to radiation from an unshielded source. This harm is a deterministic effect, meaning that the health effect is observable and directly related to the ionizing radiation received by an individual person and is associated with higher doses of radiation. By contrast, a stochastic health effect is related to low doses of ionizing radiation and is probabilistic in that a particular individual may not manifest a health effect but a large population of exposed individuals would have a certain fraction showing effects. The fraction depends on the dose received by the population. In addition to not accounting for these stochastic (probabilistic) health effects, the IAEA's categorization scheme does not factor in economic and social disruption

BOX 7-2
IAEA Categories for Radiological Sources

Category 1 has the safety concern that an unshielded source would likely cause permanent injury to someone who was in close contact for more than a few minutes and could be fatal for contact beyond several minutes to an hour. The thresholds of radioactivity for radionuclides in Category 1 sources are 60 Terabecquerel (TBq) for americium-241, 30 TBq for cobalt-60, 100 TBq for cesium-137, and 80 TBq for iridium-192. Examples of Category 1 sources are radioisotope thermoelectric generators; panoramic irradiators used in sterilization applications; and large self-shielded irradiators used in blood and research irradiation, teletherapy, and stereotactic radiosurgery devices.

Category 2 sources also pose a concern due to the potential to cause permanent injury to someone in contact with an unshielded source for many minutes to an hour and are possibly fatal when in contact for hours to days. The corresponding thresholds of radioactivity content are 0.6 TBq for americium-241, 0.3 TBq for cobalt-60, 1.0 TBq for cesium-137, and 0.8 TBq for iridium-192. Examples include smaller self-shielded irradiators, industrial gamma radiography, well-logging devices, and calibrators. These first two categories clearly can pose a health threat and thus need strong security protection as well as safety protection.

Category 3 sources are in the middle ground of unlikely to be fatal from the radiation from one unshielded source but also cross over to posing health concerns if enough Category 3 sources are aggregated to cross over to the Category 2 level. The thresholds of radioactivity content for Category 3 sources are 0.06 TBq for americium-241, 0.03 TBq for cobalt-60, 0.1 TBq for cesium-137, and 0.08 TBq for iridium-192. Examples include high- and medium-dose-rate brachytherapy, fixed industrial gauges, and well-logging devices.

Category 4 and 5 sources do not contain enough radioactivity to pose significant concerns from unshielded sources. Examples include low-dose-rate brachytherapy, thickness gauges, portable gauges, bone densitometers, and smoke detectors.

effects. In light of these gaps, the 2021 National Academies (National Academies of Sciences 2021) study highlighted that the IAEA and government agencies responsible for regulating sources should consider reframing the categorization scheme to include stochastic, economic, and social effects.

A 2019 event illustrates that even a relatively small amount of radioactivity released from a larger source can generate substantial economic and disruptive effects. On May 2, 2019, International Isotopes, a subcontractor to the U.S. Department of Energy, was assigned to remove a Category 1 sealed cesium-137 source (107 TBq) from the Harborview Research and Training Facility in Washington State. This assignment was made under the National Nuclear Security Administration (NNSA) program to remove, secure, and replace higher-activity cesium-137 sources. The subcontractor had difficulties in removing the source and in the process inadvertently released a small amount of cesium chloride (about 37 GBq), which is less than a Category 3 amount. The resulting contamination of the building resulted in 13 workers and observers receiving low doses no greater than 0.55 mSv. However, more than 200 researchers and laboratory staff had to be relocated. The resultant disruption of more than 80 funded research programs with budgets at tens of millions of dollars led NNSA to project that the final cost for response, recovery, remediation, and reconstruction will ultimately exceed $100 million (National Academies of Sciences 2021).

7.2 NNSA'S PROGRAMS TO REDUCE RISK

> FINDING 7-2: The United States maintains a robust program across several agencies, and with international partners, to detect, counter, and respond to the possibility that a terrorist or terrorists could obtain and use radiological materials in an RDD or RED. The NNSA also has effective programs in place to support the development and deployment of alternative technologies to replace radioactive sources, taking into account the need to have cost-effective devices for maintaining beneficial services.

In the aftermath of the collapse of the Soviet Union in 1991, the risk that nuclear and radiological materials might end up in the hands of criminals or terrorists was a major concern. As a result, the United States and Russia established programs to better secure facilities that held these materials and establish detection systems at border crossings should these security controls fail and traffickers try to smuggle them out of the country.

The Material Protection Control and Accounting (MPC&A) program, whose mission was to protect these materials when they could not be eliminated, was one of the earliest DOE programs. The DOE and NNSA also established Nuclear Security Centers of Excellence in Obninsk. In 2004, partly in response to 9/11, the Global Threat Reduction Initiative (GTRI) was launched to secure, protect, and remove vulnerable nuclear and radiological materials at civilian facilities worldwide. The original structure of the GTRI for radiological security was a global regional-based approach, and a separate division was established for domestic work. That has evolved in the reorganization under Global Material Security (GMS).

7.3 CURRENT PROGRAMS TO REDUCE RADIOLOGICAL RISK

Since the 9/11 attacks, the NNSA has established a number of programs to improve the security surrounding the most significant radiological sources both in the United States and internationally.

The NNSA, within its protect, control, and respond program, has specific activities designed to develop security systems to protect facilities containing the highest-level sources. One of those activities, launched in 2021, is the RadSecure 100 radiological security initiative. The objectives are to remove radiological material from facilities where feasible and improve security at the remaining facilities located in 100 metropolitan areas throughout the United States (National Nuclear Security Administration 2021).

The NNSA's focal point for its radiological security programs is the Office of Radiological Security (ORS). The ORS has a three-pillar strategy: (1) protection of radioactive sources in medical, research, and commercial use; (2) removal and disposition of disused sources; and (3) reduction in use of sources by promoting adoption and development of non-radioisotopic alternative technologies. For the first pillar, the ORS works with partner agencies, states, local governments, and tribal nations to help implement security requirements (see more details below) through development and deployment of hardening of devices containing sources, alarming buildings, training law enforcement personnel, and tracking technologies during transportation of sources. The ORS leverages the capabilities of national laboratories in these efforts with Pacific Northwest National Laboratory and Sandia National Laboratories assigned as lead laboratories. In addition, for implementing the second pillar, Los Alamos National Laboratory and Idaho National Laboratory have helped remove and secure thousands of disused and excess sources.

For the third pillar, the ORS has enlisted the national laboratories network to help develop technologies, and it provides R&D funding to companies to create alternative technologies that could replace sources. Another funding approach is to support analysis of studies that could show that an alternative technology can provide equivalent results to a radioactive source.

In 2014, the ORS began a highly successful program to replace cesium-137 irradiators with alternative technologies. This Cesium Irradiator Replacement Project (CIRP) offers incentives for users of these irradiators to switch to alternative technologies such as X-ray. Incentives include covering the removal and disposal costs and providing cost-share (typically 50 percent) for the purchase of the alternative. This has proven to be a powerful incentive since the disposal cost for a cesium-137 irradiator is on the order of $300,000. CIRP also includes cobalt-60 sources used in irradiators. In 2015, about 750 cesium irradiators (420 blood irradiators and 330 research irradiators) and 100 cobalt-60 (20 blood and 80 research) were identified in the United States as potentially eligible for CIRP. The 2019 National Defense Authorization Act (Section 3141) established the goal of replacing all cesium-137 blood irradiators with X-ray devices by December 31, 2027, via CIRP.

Internationally, the IAEA maintains programs to assist member states in improving the security of radiological materials at facilities and transportation. It has programs to "repatriate" disused sources that are no longer needed but still pose a risk (International Atomic Energy Agency 2023). The IAEA, in its role of issuing advice on safety and security practices to its member states, has worked with these states and has published several guidance documents relevant to safety and security of radiological materials. Notably, the Code of Conduct on the Safety and Security of Radioactive Sources (revised and issued in 2004) gives a framework for effective safety and security practices throughout the lifecycle of radioactive sources (International Atomic Energy Agency 2004). While this code is not legally binding, member states are encouraged to implement its practices. Some experts such as Ambassador Kenneth Brill have called for making the code legally binding "to promote national accountability" in international efforts to prevent radiological terrorism (Brill and Bernhard 2020). One of the practices in the code is to track radioactive sources throughout their lifecycle; how the United States has implemented this tracking is detailed below.

The U.S. Nuclear Regulatory Commission (NRC) has responsibility for licensing and regulating civilian use of radiological materials in the United States. The regulations cover both safety and security of these materials. Regulations in 10 CFR Part 20 titled "Standards for Protection against Radiation," Subpart I, "Storage and Control of Licensed Material" include security requirements for all radiological materials unless specifically exempted. In response to security concerns following the 9/11 terrorist attacks, the NRC issued orders in November 2005 for licensees to provide additional security for Category 1 and 2 sources. These orders were replaced by formal regulations in 2013. These regulations in 10 CFR Part 37 titled "Physical Protection of Category 1 and 2 Quantities of Radioactive Material" specify "requirements for physical security, source monitoring, personnel background checks, facility security plan, local law enforcement protection, training, and documentation" (National Academies of Sciences 2021). Notably, Part 37 only applies to Category 1 and 2 sources, as defined by the IAEA categorization scheme. Part 37 covers lower category sources if their aggregate amounts at a facility meet or exceed the Category 2 threshold. As of 2022, 39 of the 50 U.S. states belong to the Organization of Agreement States. As such, they regulate radiological materials within their states and must meet at a minimum the NRC's regulatory requirements and may promulgate stricter oversight of certain radiological materials.

The NRC has implemented the National Source Tracking System (NSTS) that serves as a national registry of all Category 1 and 2 sources used in the United States. The sources in the NSTS are organized as discrete sources and not by device or use. Certain devices have more than one discrete source; for example, sterilization devices can have more than hundreds of cobalt-60 sources. The NRC requires licensees to update the NSTS when they transfer a source or sources to another licensee or out of the country. The NSTS is a useful mechanism for providing an understanding of the lifespan of sources within the United States.

The U.S. Department of Transportation (DOT) coordinates with the NRC and the U.S. Department of Homeland Security in regulating the safe and secure transport of radiological materials in the United States. The DOT and NRC have a memorandum of understanding that details roles in package review, inspection, reporting of accidents and other events, and information sharing. The DOT has approved specific containers for certain types of radioactive sources.

7.4 THE EVOLVING SECURITY LANDSCAPE

FINDING 7-3: The security landscape for preventing the development and use of an RDD or RED continues to be challenging. There is a growing opportunity for RDD or RED attacks due to both the increased use of radiological materials and relevant emerging technologies. At the same time, these technologies may also assist the U.S.-led efforts to address the threat posed by an RDD or RED. Led by the NNSA's

Office of Radiological Security, the United States has a strong awareness and has instituted programs to understand both the risks and benefits of new technologies.

As discussed earlier in this report, violence by far-right extremist groups has risen globally and domestically in recent years. Evidence suggests that these groups have considered and continue to have interest in the use of chemical, biological, radiological and nuclear (CBRN) weapons. One report (Fleer 2020) considered three incidents involving radiological materials. Two that received considerable coverage include the 2017 incident where Brandon Russell, the founder of Atomwaffen, was found to be in possession of explosive materials as well as radiological materials. In another case, James G. Cummings was found to have a cache of radiological materials in his home suitable for building a dirty bomb (Fleer 2020).

Concurrently, there has been the rapid development and global dispersions of new technologies that have implications for managing the RDD and RED threat. These technological developments include additive manufacturing, artificial intelligence/machine learning, quantum computing, 5G networking, Internet of Things, autonomous systems and vehicles, commercial satellite imagery. Individually and in combination, these new technologies pose both risk and opportunity; that is, they have the potential to improve U.S. capabilities to detect adversary actions and, alternatively, could be exploited by adversaries. To this end, the DOE/NNSA's Strategic Outlook Initiative has a pilot enterprise-wide analytical effort underway for examining "over-the-horizon" technological developments that may impact DOE/NNSA's mission (National Nuclear Security Administration and U.S. Department of Energy 2021).

7.5 UPDATE ON RADIOACTIVE SOURCE REPLACEMENTS

An important approach to reducing the RDD and RED threat is to replace widely used commercial sources with alternative technologies. A major effort has been to phase out the use of high-risk cesium-137 sources, particularly in blood irradiators where X-ray technology offers an affordable replacement technology. The National Research Council (now the National Academies) made this recommendation in 2008 in a consensus study report, and the NNSA has since instituted programs such as CIRP, as described above, to implement that recommendation (National Research Council 2008).

Similarly, there have been a number of recommendations to secure facilities with teletherapy tools containing cobalt-60, redesign those tools to build in enhanced safeguards, or find alternatives to the use of cobalt-60 (National Academies of Sciences 2021). With regard to sources that are disused and have reached their end of life, the NNSA has responded to the recommendations of the National Academies and the IAEA and developed programs with international partners to provide means to obtain and secure such sources (International Atomic Energy Agency 2023; U.S. Department of Energy 2023).

One important barrier to radioactive sources replacement efforts is that disposal costs for disused sources can be expensive especially for higher-activity disused sources, and disposal facilities for these sources may not be available in many countries. In addition to known and accountable disused sources, orphan sources pose challenges because these sources are by definition outside of regulatory control and accounting systems. Thus, orphan sources are particularly vulnerable to theft or diversion to malicious non-state actors. More efforts are required to implement better regulatory and accounting systems in countries across the globe to identify and eliminate orphan sources. The IAEA has guidance on how to implement effective regulatory and accounting systems. The NRC via its international program office can also provide guidance to other countries, and the NRC can serve as a role model. It is also important to invest in efforts to procure and safely dispose of orphan sources.

RECOMMENDATION 7-1: The United States—with the NNSA as the lead, and in cooperation and partnership with the IAEA and other international organizations—should strengthen and accelerate current national and international activities and programs for end-of-life management of sources. Such efforts should identify disused and orphan sources and ensure that there are financial guarantees for safe and secure disposal of such materials as mentioned in a previous National Academies study (National Academies of Sciences 2021).

RECOMMENDATION 7-2: The United States—with NNSA as the lead, and in cooperation and partnership with industry—should continue and, where feasible, expand its efforts to phase out high-risk cesium-137 and cobalt-60 sources by developing and deploying reliable alternative technologies such as X-ray irradiators. Where replacement is not feasible, the NNSA should continue to assess the security risks of facilities and develop security systems to reduce the risks attendant with cesium-137 and cobalt-60.

During its information gathering, the committee reviewed the recommendations contained in an earlier National Academies report titled *Radioactive Sources: Applications and Alternative Technologies* (National Academies of Sciences 2021) and agrees that several actions suggested for implementation in that report have merit and would complement ongoing actions to enhance the security of radiological materials. These include:

a. Prioritize research funding and development of alternatives for alternatives to the use of radiological materials, where no such alternatives exist but would be beneficial.
b. In low- and middle-income countries where there are logistical or other barriers to deployment of alternative technologies, focus on ensuring security of radioactive sources already in use, while engaging in cooperation with such countries to address these barriers, where possible.
c. Support equivalency studies for applications such as oil-well logging, research irradiation, and radiography to provide a technical basis for development of potential alternatives.

The committee also noted that the prior report recommended measures related to reframing radiological source characterization schemes, domestically and internationally, to account for economic and social impacts, in addition to any deterministic effects of ionizing radiations from these sources and to extend source tracking systems for category 1 and 2 sources, to include category 3 sources. To date, such proposals have failed to secure regulatory approval with any national regulatory body, principally due to the rigors of regulatory cost-benefit analyses and the challenges of regulating to public confidence and perceptions of risk. Although the committee did not hold a uniform view on adoption of these measures, the committee acknowledges the challenge of regulating to perceived levels of risk with most members seeing the merit of future consideration of source tracking measures, as changing security circumstances may warrant.

References

Brill, Kenneth C., and John H. Bernhard. 2020. "Preventing the Preventable: Strengthening International Controls to Thwart Radiological Terrorism." *Bulletin of the Atomic Scientists* 76 (4): 206–209. https://doi.org/10.1080/00963402.2020.1778371.

Fleer, BreAnne K. 2020. "Radiological-Weapons Threats: Case Studies from the Extreme Right." *The Nonproliferation Review* 27 (1–3): 225–242. https://doi.org/10.1080/10736700.2020.1775987.

International Atomic Energy Agency. 1988. *The Radiological Accident in Goiânia*. Vienna: IAEA. https://www-pub.iaea.org/mtcd/publications/pdf/pub815_web.pdf

---. 2004. "Code of Conduct on the Safety and Security of Radioactive Sources." https://www.iaea.org/publications/6956/code-of-conduct-on-the-safety-and-security-of-radioactive-sources

---. 2005. "Categorization of Radioactive Sources." https://www.iaea.org/publications/7237/categorization-of-radioactive-sources

---. 2023. "Disused Sources." https://www.iaea.org/topics/disused-sources.

Jefferson Lab Resources. 2023. "The Element Cobalt." It's Elemental. https://education.jlab.org/itselemental/ele027.html.

National Academies of Sciences, Engineering, Medicine. 2021. *Radioactive Sources: Applications and Alternative Technologies*. Washington, DC: The National Academies Press.

National Nuclear Security Administration. 2021. "NNSA Launches RadSecure 100 Radiological Security Initiative in 100 U.S. Cities." https://www.energy.gov/nnsa/articles/nnsa-launches-radsecure-100-radiological-security-initiative-100-us-citiesNational Nuclear Security Administration, and U.S. Department of Energy. 2021. "Prevent, Counter, and Respond—NNSA's Plan to Reduce Global Nuclear Threats FY 2022-FY 2026." https://www.energy.gov/nnsa/articles/prevent-counter-and-respond-strategic-plan-reduce-global-nuclear-threats-npcr

National Research Council. 2008. *Radiation Source Use and Replacement: Abbreviated Version*. Washington, DC: The National Academies Press.

Oncology Medical Physics. 2023. "Medical Physics Made Easy." https://oncologymedicalphysics.com/.

U.S. Department of Energy. 2023. "ENERGY.GOV." https://www.energy.gov/.

FIGURE 8-1 A mobile radiation detection and identification system (MRDIS, orange structure in upper and lower images) allows vehicles with containers to pass through and be scanned for radioactive signatures (lower image). Photos are from National Nuclear Security Administration project at the port of Salalah in Oman where cargo is offloaded from large ships into smaller vessels appropriate for travel through the Suez Canal or taken off-site via trucks. The project was a multi-lab project. Sandia was the system lead, Los Alamos National Laboraotry developed the sodium iodide collectors, and Pacific Northwest National Laboratory assisted in logistics. This is an example of deploying technologies to improve the visibility and accountability of containerized cargo to both deter trafficking of nuclear materials across borders and improve the safety of commercial shipping.
SOURCE: Photos courtesy of Dr. Rodney K. Wilson.

8

Detection and Interdiction Efforts within and Outside the Global Supply System

Arms control and nonproliferation treaties and agreements and other efforts to bolster the security of nuclear weapons and nuclear materials outlined in earlier chapters have made important contributions to reducing the nuclear terrorism risk. With fewer weapons and enhanced transportation and storage safeguards for nuclear materials, there are fewer potential opportunities for non-state actors to gain access to the means for carrying out a nuclear terrorist attack. However, as chapters 6 and 7 have outlined, there is still much work to be done to minimize and secure weapons-usable fissile and radiological materials. If the security measures and safeguards currently in place are compromised, weapons and nuclear materials might end up in the possession of those who seek to move them or use them. This potentiality translates into the need for enhancing capabilities to detect and intercept weapons or nuclear materials before they get into the hands of a terrorist or, if not then, before a nuclear device reaches its intended target—a layered approach to defense. This risk places a premium on developing and sustaining robust international and domestic counter-smuggling capabilities within the legitimate global trade and transportation system and along traditional and new smuggling routes outside those systems.

8.1 THE LONG-STANDING CHALLENGE OF POLICING THE GLOBAL SUPPLY CHAIN

The Cold War was followed by a surge in global economic activity, fueled by economic liberalization policies and free trade agreements to ease the movement of people,

BOX 8-1
Summary

Non-state actors can move nuclear weapons, materials, and equipment by exploiting well-established criminal pathways for smuggling. This is true even in the face of the many detection and interdiction measures put in place since 9/11. Opportunities exist to enhance supply chain transparency and accountability by strengthening industry partnerships and taking advantage of improvements in technologies to include artificial intelligence and machine learning. Within the global supply system, these technologies can expand the means for identifying anomalies and dangerous materials hidden within legitimate shipments. They can also help provide rapid forensics that can support incident response and recovery. Concurrently, strengthening efforts to counter cross-border smuggling outside the legitimate trade and travel routes also remains critical for managing the nuclear terrorism risk.

Highlights

- Efforts to detect and intercept nuclear materials within the global supply system and along traditional and nontraditional smuggling routes require increased attention.
- The consequences of a nuclear terrorist incident that both exploits and targets the global supply system could be devastating, as demonstrated by the fragility of the supply chains during the pandemic.
- The efforts to detect radiological and nuclear material entering the United States remain almost exclusively at international ports of entry, even with criminal and terrorist organizations moving other materials outside of these entry points along traditional and new smuggling routes.

conveyances, and goods across national borders.[1] This led to a dramatic expansion in the variety and volume of goods circulating around the world and the rapid expansion of the cargo and transportation networks that facilitated that growth.[2] Intermodal logistics were revolutionized, dramatically lowering the cost of operating complex global supply chains. Increasingly, companies realized that they could dispense with the expense of maintaining large inventories in warehouses or in the backrooms of department

[1] Prominent examples of cross-border liberalization efforts include the 1994 North American Free Trade Agreement (NAFTA) between the United States, Canada, and Mexico, and the Schengen Convention of 1990 that launched the end of internal border controls among the now 27 European countries who are party to the agreement.

[2] The number of containers handled by the world's port terminals in 1993 doubled by 1998 and doubled again by 2001. Larger vessels were constructed to meet this demand, and, as the ships grew in size, so too did the port facilities to handle them. In the mid-1970s, the typical growing container ship carried 1,500 twenty-foot-equivalent units (TEUs). Today, the latest ships carry more than 20,000 TEU from "mega-ports" such as Singapore, Rotterdam, Dubai, Hong Kong, and Shanghai.

stores. They instead relied on "just-in-time shipping" where the transportation system effectively served as a mobile warehouse. As these companies grew, so did their global transportation needs, with the biggest companies acquiring the leverage to lower shipping rates. The small profit margins generated as a result increased the pressure on the transportation industry to reduce costs by achieving greater economies of scale and efficiency in operations.

One outcome of the speed with which the global supply system evolved in the 1990s was that security measures within the maritime, aviation, and surface transportation sectors did not keep pace. Criminals took advantage of this situation. Given the limited capacity to inspect the huge number of conveyances and containers crossing borders and transiting through ports and airports each day, traffickers found that there were ample opportunities to smuggle all forms of contraband, ranging from stolen vehicles and illicit narcotics to small arms and counterfeit goods.

In the aftermath of 9/11, efforts to bolster the security of the system had to overcome three challenges: (1) the complexity and multiplicity of actors involved with an increasingly globalized system, (2) uneven oversight of the system, both internationally and among agencies, with no overall lead, and (3) private sector cost-burden concerns associated with any new security measures.

8.2 EFFORTS TO IMPROVE DETECTION AND INTERDICTION

FINDING 8-1: Post-9/11 efforts to improve transportation, cargo, and border security have significant limitations but provide a strong foundation for enhancing industry and international partnerships. Such partnerships are required to effectively deter and counter nuclear and radiological materials smuggling within the global supply system.

Concerns about the poor state of transportation, cargo, and border security received significant attention after September 11, 2001. The U.S. national security community's guard had been down, in part due to what the 9/11 Commission would later call a "failure of imagination." After the attacks on the World Trade Center and the Pentagon, attention was directed to how terrorist groups might exploit other transportation conveyances, including vessels, cargo planes, trains, trucks, and intermodal containers. Multiple federal agencies embarked on stepped-up efforts to bolster transportation and cargo security, but no single department or agency had overarching responsibility for security.

The current array of transportation, cargo, and border security measures were largely developed in the George W. Bush Administration with the most prominent measures including:

- **The International Ship and Port Facility Security (ISPS) code**
  In the fall of 2001, the U.S. Coast Guard (USCG) worked through the International Maritime Organization (IMO), a specialized agency of the United Nations, headquartered in London, to establish the International Ship and Port Facility Security code (U.S. Coast Guard 2014). The ISPS code went into effect on July

1, 2004, establishing minimum-security requirements for vessels, shipboard personnel, and port facilities to "detect security threats and take preventative measures against security incidents affecting ships or port facilities used in international trade."

- **Customs-Trade Partnership Against Terrorism (CTPAT) and the Container Security Initiative (CSI)**
  The U.S. Customs Service (later Customs and Border Protection or CBP) launched the *Customs-Trade Partnership Against Terrorism* (U.S. Customs and Border Protection 2023) followed in 2002 by the *Container Security Initiative* (CSI). The CTPAT identified supply chain security "best practices" that companies involved with importing goods into the United States were encouraged to adopt voluntarily. The CSI involved foreign customs officials agreeing to host CBP inspectors and collaborate on inspecting U.S.-bound containers identified as high risk at ports of loading.

- **Second Line of Defense and Megaports Programs (currently the Nuclear Smuggling Detection and Deterrence Program)**
  After 9/11, the National Nuclear Security Administration (NNSA) provided direct assistance to strengthen the technological capabilities of partner countries to combat radiological and nuclear materials smuggling. In 2003, the NNSA launched the Megaports Initiative (Clarke 2009) providing radiation detection equipment, training, and technical support to foreign customs and overseas port authorities and terminal operators. In 2005, the Domestic Nuclear Detection Office was established within the U.S. Department of Homeland Security (DHS) and assigned responsibility to manage the nuclear terrorism risk at U.S. border crossings and within U.S. ports of entry.

- **Proliferation Security Initiative**
  The *Proliferation Security Initiative* (PSI) was launched in May 2003, enlisting countries to endorse the PSI Statement of Interdiction Principles "which commit participants to establish a more coordinated and effective basis through which to impede and stop WMD, their delivery systems, and related items" (U.S. Department of State 2023a). The PSI countries commit to interdict transfers to and from states and non-states of proliferation concern, develop procedures to facilitate exchange, strengthen national legal authorities to facilitate interdiction, and take specific actions in support of interdiction efforts (U.S. Department of State 2023b). To date, 107 countries have endorsed these PSI principles.

- **United Nations Security Council Resolution 1540**
  The U.N. Security Council passed Resolution 1540 in 2004, establishing the 1540 Committee to provide legislative guidance and technical support to U.N. member states who agree to "refrain from providing any form of support to non-State actors to develop, acquire, manufacture, possess, transport, transfer

or use nuclear, chemical or biological weapons and their means of delivery, in particular for terrorist purposes." The resolution has been reaffirmed on nine occasions, most recently on November 30, 2022, for an additional 10-year period.

Notwithstanding the good work of the departments and agencies and these post-9/11 programs, there are gaps in the overall security picture. The ISPS code addresses the security of the ships, terminals, and personnel but not the cargo they handle and transport. The U.S. Coast Guard's International Port Security (IPS) program, which is responsible for confirming compliance with the ISPS Code, has never been adequately funded to carry out regular inspections at the individual port facility level. The CTPAT currently has more than 11,000 certified partners that account for more than 50 percent of cargo (U.S. Customs and Border Protection 2023), by value, imported into the United States, but CBP does not have adequate staffing to validate that the security measures of CTPAT members are reliable, accurate, and effective. They are not able to conduct periodic audits of CTPAT-certified partners. This makes it difficult to distinguish between those companies that are making good-faith efforts to implement supply chain security best practices and those that are not. The CBP's Container Security Initiative is operating in 61 ports and prescreens cargo manifests and other trade information for more than 80 percent of U.S.-bound maritime containerized cargo (U.S. Customs and Border Protection 2022). Only a fraction of one percent of containers, however, are subject to nonintrusive inspection at the overseas ports of loading. Upon arrival in a U.S. port, CBP typically inspects just 3% of inbound cargo containers (Green Worldwide Logistics 2019). The selection of which containers warrant such an inspection is based primarily on an algorithm developed to identify high-risk shipments. In 2021, however, when the National Cargo Bureau conducted inspections of 500 containers that are not traditionally inspected, they found that 55 percent of containers were out of compliance with safeguard regulations and 2.5% of inspected dangerous goods containers were found to include misdeclared cargo that "represented a serious risk to crew, vessel, and the environment" (National Cargo Bureau 2020). Nine major containership fires reported in 2019 were attributed to poorly stowed, undeclared, or misdeclared dangerous cargo (National Cargo Bureau 2020).

The post-9/11 programs to counter nuclear terrorism and proliferation launched by the U.S. Department of Energy, Department of Defense, and Department of State also have significant limitations. In the absence of nonintrusive inspection equipment that can identify shielding, the stand-alone radiation portals that have been deployed at major seaports and border crossings under the U.S. Department of Energy's Second Line of Defense, MegaPorts, and successor programs are not able to detect shielded radiological materials. Thus, criminals or terrorists could use readily available materials such as lead to encase nuclear materials or a weapon that might then be able to pass through a radiation portal without setting off an alarm.

Interdiction efforts of suspected nuclear shipments that are pursued under the Proliferation Security Initiative face the practical challenge of gaining access to the contents of individual containerized cargo shipments once they are loaded aboard a container ship. When containers are stowed, the space between them typically is only 18–24 inches. The containers are placed in stacks that can be 10 or more deep below and above decks, and as many as 20 across. A boarding team has no practical way to gain access to an individual container while it is aboard a vessel. Instead, the ship must come into a port equipped with a gantry crane to remove the containers stacked on top of and around the suspected container. This may mean the ship must divert to a port where the container can be safely offloaded. Consent of the state that has registered the ship (flag state consent) or other permissions such as from the ship's master or owner must first be obtained. Consideration must also be given to liability if the cargo ship is significantly delayed from its schedule.

The U.N. Security Council Resolution 1540 (UNSCR 1540) has played a helpful role in advancing the norm that U.N. member states should independently and collectively work to reduce the nuclear terrorism risk. It calls for each state to prepare a national implementation action plan, but these are done on a voluntary basis and not all states have done so. The 1540 Committee can respond to requests for technical assistance, but the Committee is not authorized to confirm compliance by member states. With respect to preventing shipments of nuclear and other materials, the Committee has not yet developed programs to guide member states on what they should be doing to mitigate the risk of non-state actors transporting nuclear materials within their jurisdictions and across their national borders. Nor has the Committee established a collaborative relationship with its fellow U.N. agency, the IMO, to undertake counterproliferation efforts in seaports and within global shipping channels. Today, many developing countries simply lack the resources and capabilities to prevent the transport of nuclear materials within and across their borders.

To summarize, the authorities and programs that touch on ports, ships, cargo, counterterrorism, and counterproliferation are spread across an array of U.S. departments and agencies with varying levels of domestic and international reach. Individually and collectively, these efforts have raised awareness, helped to advance global norms, engaged international partners, facilitated closer cooperation with the private sector, and provided expanded capacity for detection and interception of illicit nuclear materials. Yet, all these efforts have been advanced in an uncoordinated manner with uneven funding and staffing support to sustain them. No one agency or department has been assigned to serve as the overall lead for detection and interdiction efforts.[3]

[3] The one attempt to develop a comprehensive approach to global supply chain security occurred during the Obama Administration. After a two-year interagency process, the result was a strategy document (The White House 2012) that only set very general goals such as calling for the integration of federal efforts and enhancing coordination with the international community. The strategy did not provide for or lead to any modifications or additions to the many programs launched in the aftermath of 9/11 to prevent the movement of nuclear materials, technology, and expertise to hostile states and terrorist organizations.

8.3 THE RISK NUCLEAR TERRORISM POSES TO THE GLOBAL SUPPLY CHAIN

FINDING 8-2: A nuclear incident involving the global supply system would expose gaps in the system's security and lead to catastrophic economic consequences arising from system-wide delays while new security measures were developed and deployed.

Highly dangerous materials continue to evade safeguards within the maritime transportation system, on which the smooth operation of global trade flows depend. (National Cargo Bureau 2020) Should a terrorist organization decide to put the current security measures to the test by intercepting a container from a "trusted shipper" and inserting a shielded radiological dispersal device (RDD), it would be very difficult to detect the shipment in the absence of an intelligence tip. If the RDD were detonated at an arrival port, the efficacy of all the post-9/11 port and container security measures would be called into question.[4]

Beyond the direct damage done by the nuclear device itself, the aftermath of a nuclear incident would have widespread economic consequences. These consequences would arise from the inevitable public anxiety that the incident would generate about the dangers posed by uninspected cargo containers. Addressing this anxiety by physically inspecting all inbound cargo containers would lead to supply chain gridlock. These inspections could not be done aboard a loaded vessel at anchor or at sea, and suspect vessels might not even be allowed to dock and unload uninspected containers so that they could be examined within a port. Under the "Implementing Recommendations of the 9/11 Commission Act" of 2007, there is already a legal requirement mandating that 100 percent of U.S.-bound cargo containers undergo nonintrusive imaging and pass through radiation-detection equipment prior to their being loaded overseas. This requirement has been waived by the secretary of homeland security at two-year intervals since 2007, but there would likely be enormous public and political pressure to immediately implement the law in the aftermath of a terrorist incident. One hundred percent inspections would result in the kind of vessel backups that took place during the COVID-19 emergency, with the associated cascading global supply chain effects and impacts on worldwide economic activity.

---

[4] By compromising a "trusted shipper" to send a compromised shipment to the United States, a terrorist organization would expose the limitations of the Customs-Trade Partnership Against Terrorism (CTPAT) safeguards, the International Ship and Port Facility Security (ISPS) code, and the ability to CBP to target a high-risk container at an overseas port under the Container Security Initiative (CSI) protocol. If shielded, the compromised shipment is likely to have evaded any radiation technology equipment deployed at the original and arrival ports.

8.4 ENHANCING THE MEANS TO MONITOR CARGO SHIPMENTS

FINDING 8-3: Technologies are available to enhance supply chain transparency and the means to detect contraband including nuclear materials. These improved methods include nonintrusive inspection technologies and processing scanned images with the assistance of AI and machine learning to better detect and intercept contraband to include nuclear materials.

It is possible to adapt port facility operations to use nonintrusive inspection (NII) technology to routinely scan all containers entering a port facility so as to confirm that the contents do not pose a nuclear or radioactive risk to the terminal, ship, and its crew (Bakshi, Flynn, and Gans 2011).[5] Note that scanning for nuclear materials is part of the overall contraband identification process, along with other types of contraband. When a container triggers an alarm, it can be transferred to a secondary inspection area to scan the contents by more sensitive NII equipment. In most instances, this more detailed examination would resolve the concern in minutes[6] and the container could then be cleared and transferred back into the container yard in time to make its scheduled voyage. This additional scanning data could then be forwarded to customs inspectors in

[5] When it comes to integrating NII equipment into port operations, terminal managers are in the best position to address the operations management and system engineering issues. Embedding drive-through portals into the terminal gate structure is relatively straightforward. Placing the equipment in the container yard or quayside to support the scanning of transshipment containers is a more complicated traffic management challenge but not an insurmountable one since the images can be typically collected in under a minute.

The logic of making this fundamental shift—from a default of no inspection to a default of inspection—derives from the fact that in this case it is easier to prove a negative than a positive. That is, in order for a container to pose a risk of nuclear terrorism, it would have to contain both radiological material and shielding to prevent detection of that material, so the goal is to determine that neither is present. A radiation portal monitor can determine the presence of radiation, but the contents of the container would need to be scanned in order to identify heavy metals with sufficient density that can be provided by shielding to defeat radiation detectors. These metals would need to be lead or another element with a high atomic number, generally referred to as "high-Z" materials. Accordingly, the use of NII could be largely automated. If a container driven through a radiation detection and scanning portal had neither radioactivity or high-Z materials—and the overwhelming majority of containers do not—it could be automatically cleared to be stored in the container yard or transferred directly for loading aboard a vessel.

[6] NII technology continues to improve to include new passive system technology first invented at Los Alamos National Laboratory. The current version allows for automated alerts on radiological material, material discrimination based on density and automated material identification alerts, and a machine-learning library that can support the continued refinement of algorithms to accurately interpret images. The Generation 3 Multi-Mode Passive Detection System (MMPDS Gen3) is a product of Decision Sciences International Corporation (Decision Sciences 2023).

the destination port to supplement their information.[7] In the rare instances where alarms are not be resolved by the secondary inspection scanning, the appropriate protocol would be for officials in both the loading and destination ports to be alerted[8] and the container to be moved to a secure holding area where its contents could be inspected by local officials or in collaboration with the Container Security Initiative team that the United States has deployed overseas.[9] Any breaking of the container seal to gain physical assess to the container's contents would only be done by authorized inspectors.

---

[7] It is the ability to use NII data to "pull bits" instead of "pulling boxes" that can make it cost-effective. In the first comprehensive analytical and technical assessment of the operational impact of container inspections in international ports, a 2010 study collected detailed data on the movement of more than 900,000 individual containers at two of the world's largest international container terminals. (Bakshi, Flynn, and Gans 2011; Finklea 2020).

Since containers typically arrive two to three days before their voyage, by the time they are identified for inspection by U.S. customs officials, they are almost always already sitting in a stack in the container yard, waiting to be loaded on a container ship. Containers are typically stacked up to six high in most major ports. This translates into the need to lift and move out of the way the containers on top of a targeted container in order to transport it. Then the container must be placed on drayage to be carried to the customs inspection facility, await scanning, and then be transported back to the stack. The project used these records as the basis for a simulation analysis that estimated the effect of a number of inspection protocols on terminal operations. It determined that automatically scanning all containers upon arrival would be more operationally efficient and cost-effective than conducting targeted pre-loading inspections of a very small percentage of U.S.-bound containers. It was calculated that the cost of these inspections would average $110 each and could create a significant backlog at the inspection facility if overseas officials were directed to inspect as little as five percent of U.S.-bound cargo at any given time using the Container Security Initiative protocol. Alternatively, the study found that automatically scanning all containers upon arrival could be covered by a $15 per container Terminal Security Charge.

[8] Another way to minimize the impact on cargo handling by port-of-loading inspections is to have inspections that support the interception of contraband such as drugs, currency, or counterfeit goods be done at the port of arrival. Contraband that does not pose a direct threat to the safety of the terminal or the vessel transporting it does not need to be interdicted before loading. Furthermore, the laws defining contraband are not universal, which translates into goods being potentially legal in the exporting jurisdiction even though they are illegal in the importing jurisdiction. Accordingly, the appropriate locus of non-nuclear contraband enforcement is at the port of arrival where customs officials can use the risk-management tools normally available to them, augmented by the additional data provided by the nonintrusive image captured at the port of loading.

[9] Beyond deploying equipment to routinely collect images of a container's content, port operators should also put in place the secure data management processes that can support the automatic transfer of NII data to officials who may be interested in reviewing it. The objective should be for this data to be shared as soon as it is collected. Rapid sharing assures that government agents can exercise oversight of the port operators or (more likely) the bonded third-party entities that operators contract with to manage the on-the-ground container screening process. Having direct access to the data would also allow government inspectors to examine images of cargo in advance of loading that they have determined might pose a high risk. In this way, they could resolve their concerns without needing to alert the port operator or even the local government. In the case of contraband, they may also decide to allow the container to move through the supply system unmolested to gather intelligence and secure evidence of trafficking without alerting the criminal conspirators.

8.5 DETECTION AND INTERDICTION AT PORTS OF ENTRY

FINDING 8-4: The efforts to detect radiological and nuclear material entering the United States remain almost exclusively at official ports of entry despite the ongoing risk that criminal and terrorist organizations may move this material along traditional and new smuggling routes.

Detection and interdiction efforts at official ports of entry and commercial ports remain the nearly exclusive focus of current efforts to find potential inbound radiological and nuclear materials. This is true even though small groups or even a single individual could transport IND- and RDD-relevant materials. Personal smuggling would probably appeal to terrorists who want to retain direct possession of the material to construct and use a weapon within the United States. Therefore, illegal means of entering the country at a location between official ports of entry represents a threat vector that deserves stepped-up attention.

The difficulty of comprehensively monitoring the vast and diverse national boundaries between official ports of entry is manifold and well known. For the United States, much of these "frontier" regions are remote and characterized by rugged, difficult-to-navigate terrain where law-enforcement staffing is very sparse. Despite these challenges, there are long-standing efforts to prevent drug and human trafficking and other contraband along illicit maritime, air, and land transit routes and as a part of border control efforts outside legal ports of entry. These counter-smuggling efforts position federal, state, local, and tribal law enforcement agents to play a frontline role in nuclear counterterrorism efforts. To be successful, agents would need ongoing training and access to specialized equipment to allow them to identify and safely handle radiological and nuclear materials.

RECOMMENDATION 8-1: The United States should lead an international effort to enhance security across all elements of the global supply system by building on the post-9/11 transportation and cargo security programs and deepening international and private industry cooperation. Agencies and organizations involved with this effort should include the United Nations 1540 Committee, the IMO, DHS to include the USCG and CBP, the Department of State, NNSA, and the International Atomic Energy Agency (IAEA).

Countries and industry require uniform global standards and procedures to ensure that legitimate trade and transportation systems are not being used to move prohibited nuclear materials and contraband. The international arrangements for putting in place a system-wide approach to achieve this uniformity are largely in place.

Under the UNSCR 1540 mandate, member states are required to develop and maintain "appropriate effective border controls and law enforcement efforts to detect, prevent, and combat ... the trafficking in nuclear, chemical, or biological weapons and their means of delivery." However, without the means to routinely monitor the commercial goods and conveyances that transit through and depart from their jurisdiction, states cannot meet this obligation. The UN 1540 Committee is tasked with engaging relevant international organizations and forging effective partnerships with the private sector and industry so as "to support national and international efforts to meet the objectives of the resolution." Accordingly, the 1540 Committee would be well within its mandate to work with the IMO to incorporate new requirements into the International Ship and Port Facility (ISPS) Code. Similarly, close collaboration between the IMO and the 1540 Committee to prevent nuclear materials smuggling would be fully consistent with the maritime shipping safety and security imperative that drove creation of the ISPS Code: that is, to "establish the new international framework of measures to enhance maritime security and through which ships and port facilities can operate to detect and deter acts which threaten security in the maritime transport sector." Such cooperation would provide the 1540 Committee with a means to further member state compliance with the 1540 mandate and the IMO and would provide guidance to meet the mandates on the ISPS Code, while at the same time enhancing security at maritime borders worldwide.[10]

The U.S. DOS and the USCG should take the lead in linking the currently disconnected global counterproliferation mandate set by U.N. Security Council Resolution 1540 and the global port security requirements embedded in the International Maritime Organization's ISPS code to advance universal cargo scanning to detect prohibited nuclear materials at ports of loading. Shipping companies and marine terminal operators who directly handle most of the world's maritime containers should also be enlisted as full partners.[11]

With the technical support of the IAEA, the NNSA, and the USCG, the U.S. Department of State should work to have the IMO incorporate, within Part B of the ISPS Code, recommended guidance for uniform, performance-based standards for non-intrusive inspection (NII) and radiation detection equipment to be used in marine terminals. These agencies should also provide guidance to port industry partners on how data collected by the NII equipment can be securely shared with government inspections officials at

[10] Specifically, the guidance contained in part B of the ISPS Code should include recommended practices for ensuring that cargo entering port facilities does not pose a nuclear risk to the ships and crews transporting that cargo. This would have the constructive result of making the maritime industry a full-security partner in bolstering cargo security while simultaneously establishing common standards for the entire global maritime transportation system. The principal tenet of this part B guidance should be that port facilities should confirm, before cargo is loaded aboard a ship, that it does not possess a nuclear and radiological device or materials. This approach would also bolster forensic capabilities to more surgically identify where breaches to security may have occurred. That is, in the event of a scenario where a terrorist targets the global supply system itself as a critical infrastructure with the goal of generating mass disruption and the associate economic consequences, universal cargo scanning would support isolating the source of the attack and quickly allowing the operations of other cargo and conveyances by being able to confirm that they pose a low risk of follow-on attacks.

[11] Eight large companies handle and transport more than half of the containerized cargo worldwide to include virtually all the major ports and most smaller regional ports. These companies adopt standards of safety and security that are consistent across all their port terminal operations and that in most cases exceed the minimum standards outlined in various international instruments. Most importantly, their operational capabilities and global presence makes then best suited to cost-effectively and efficiently implement security requirements that enhance the ability to detect and intercept dangerous materials.

both the port of loading and the port of arrival if requested.[12] Artificial intelligence and machine-learning capabilities are reaching a level of maturity to routinely match cargo manifests with scanned images, and thus reduce false-positive rates.

RECOMMENDATION 8-2: The DOJ, FBI, DOE, and DHS, with support from the U.S. Department of State, should continue to deepen ongoing international law enforcement cooperation and intelligence sharing to counter nuclear smuggling efforts along illicit transit routes and between legal ports of entry. These agencies should also ensure that federal, state, local, tribal, and territorial (SLTT) law enforcement agents involved in interdiction and border control efforts receive ongoing nuclear detection training and have ready access to specialized equipment, expertise, and the means to handle radiological and nuclear materials safely.

Criminals and terrorist organizations that face enhanced security measures within the legitimate flows of global trade and the transportation conveyances will adapt by engaging in cross-border smuggling outside those flows. Countering that evolution will require more closely coordinated detection and interdiction efforts by law enforcement authorities between jurisdictions where weapons and materials may originate and the adjacent transit countries (Finklea 2020; Jancsics 2021).

References

Bakshi, Nitin, Stephen E. Flynn, and Noah Gans. 2011. "Estimating the Operational Impact of Container Inspections at International Ports." *Management Science* 57 (1): 1–20. https://doi.org/10.1287/mnsc.1100.1252. https://pubsonline.informs.org/doi/abs/10.1287/mnsc.1100.1252.

Clarke, Stephanie P. 2009. *Radiation Detection, Response & Recovery*. Edited by Office of the Second Line of Defense National Nuclear Secuirty Administration. https://www-ns.iaea.org/downloads/rw/meetings/tarragona2009/presentations-wednesday/s-clarke.pdf

Decision Sciences. 2023. "Our Product: The Next Generation of Non-intrusive Inspection." https://www.decisionsciences.com/products-solutions/.

Finklea, Kristin. 2020. *Illicit Drug Smuggling between Ports of Entry and Border Barriers*. Edited by Congressional Research Service.

Green Worldwide Logistics. 2019. "The Difference between Cargo Screening vs. Inspection." https://www.greenworldwide.com/the-difference-between-cargo-screening-vs-inspection/.

---

[12] Partnering with industry should include authorizing bonded third parties to work with customs inspectors to address and resolve alarms generated by the NII equipment when they occur. Authorization also should be provided for leveeing of a security fee for industry partners to recover the cost of their implementing these actions as a part of the authorized Terminal Security Charge that supports investments to comply with the ISPS code.

As this comprehensive approach is put in place, an interagency working group that includes NNSA, DHS's Countering Weapons of Mass Destruction Officer (CWMD), the U.S. Coast Guard, and CBP should design and conduct exercises that can support identifying gaps in the transportation and cargo security measures as well as test the response to and recovery from an incident. Industry partners should be included, facilitated by nonprofit organizations such as the National Cargo Bureau.

Jancsics, David. 2021. "Law Enforcement Corruption along the U.S. Borders." *Security Journal* 34 (1): 26–46. https://doi.org/10.1057/s41284-019-00203-8. https://doi.org/10.1057/s41284-019-00203-8.

National Cargo Bureau. 2020. "A Comprehensive Holistic Approach to Enhance Safety and Address the Carriage of Undeclared, Misdeclared and Other Non-compliant Dangerous Goods." *Safety of Life and Cargo at Sea*. https://natcargo.org/wp-content/uploads/2022/07/Holistic-Approach-For-Undeclared-Misdeclared-And-Other-Non-compliant-Dangerous-Goods_White-Paper-by-NCB.pdf.

The White House. 2012. National Strategy for Global Supply Chain Security. https://obamawhitehouse.archives.gov/sites/default/files/national_strategy_for_global_supply_chain_security.pdf

U.S. Coast Guard. 2014. "International Port Security Program–Best Practices." Last modified April 14, 2014. https://www.dco.uscg.mil/Our-Organization/Assistant-Commandant-for-Prevention-Policy-CG-5P/Inspections-Compliance-CG-5PC-/International-Domestic-Port-Assessment/International-Port-Security-Program-Best-Practices/.

U.S. Customs and Border Protection. 2022. "CSI: Container Security Initiative." Last modified October 6, 2022. https://www.cbp.gov/border-security/ports-entry/cargo-security/csi/csi-brief.

---. 2023. "CTPAT: Customs Trade Partnership against Terrorism." Last modified June 21, 2023. https://www.cbp.gov/border-security/ports-entry/cargo-security/CTPAT.

U.S. Department of State. 2023a. "Proliferation Security Initiative: Bureau of International Security and Nonproliferation." https://www.state.gov/proliferation-security-initiative/.

---. 2023b. "PSI Interdiction Principles." https://www.state.gov/psi-interdiction-principles/.

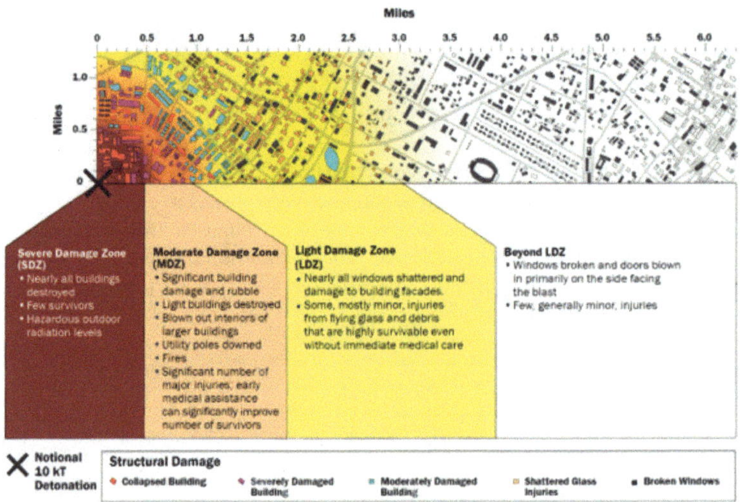

FIGURE 9-1 Blast damage zones after a 10 kT detonation. The response and recovery will be different for each zone, including degrees of assistance and timing with which first responders can arrive. Varying sizes of blast zones are shown in figure 2-1.
SOURCE: Federal Emergency Management Agency 2022.

9

Response to and Recovery from Nuclear Incidents

9.1 CURRENT RESPONSE CAPABILITIES

FINDING 9-1: Over the past decade, training capacity and response capabilities have stagnated due to insufficient funding and planning resources.

For a government to be resilient in the face of the nuclear terrorism threat against the homeland, or after a nuclear terrorist attack, preparation and planning measures must be regularly updated. Government officials at all federal, state, local, tribal and territorial levels (FSLTT) must be continuously educated about the threat landscape and their responsibilities for providing the necessary emergency management resources and capabilities.

In the wake of the 9/11 attacks, the potential for a terrorist attack involving detonation of a nuclear or radiological device without prior warning commanded a great deal of attention and concern. While state and local planning capacity for a nuclear incident after 9/11 was increased, that capacity has since deteriorated due to significant reductions in funding and planning resources (Bentz 2022). Even where there are plans in place, the number and level of exercises to test those plans and bolster local response capabilities has decreased. In the limited instances where local exercises include a radiological/nuclear scenario, the central focus tends to be on development of individual responder skills (equipment maintenance, calibration, interpretation of readings, and other technical responsibilities) rather than on how all the various parts of the response system would operate together. This translates into state and local authorities with limited exposure to the kinds of issues that large-scale exercises would facilitate such as discussions of mitigation measures, plans for cleanup, and large-scale mapping of contamination zones. It is also rare for radiological emergency exercises to incorporate the likely follow-on disruptions expected in a real emergency, nor do they consider the myriad challenges associated with recovery.

BOX 9-1
Summary

Emergency preparedness, including preparations for a nuclear or radiological event, is almost entirely dependent on local, state, and regional authorities, who have the primary responsibility to protect the population within their jurisdiction. However, in a nuclear emergency, key information, support, and rare expertise can come only from federal resources. Governors and mayors are generally not adequately trained or equipped to respond to a nuclear or radiological event. Additionally, they are confronted with competing priorities that make it difficult to devote the attention required to prepare for this kind of low-probability/high-consequence threat. The coronavirus pandemic exposed the disparate capabilities that exist across the nation's local and state jurisdictions as well as significant shortcomings in coordination among federal, state, local, and tribal authorities in an extended public health emergency. Emergency management, like the U.S. public health system, operates primarily under the purview of governors, county commissioners, and mayors for which the federal government plays a support role. For a nuclear incident, consequence management and recovery have the added burden of trying to manage an incident in the face of widespread fear. This is compounded by the dangers associated with traditional media potentially broadcasting inaccurate information in its early reporting, alongside social media providing channels for pervasive disinformation and misinformation. An adequate response to a nuclear or radiological incident requires enhanced coordination of emergency management response protocols across all levels of government and strengthened information dissemination tools for providing trusted, accurate, science-based information and techniques. Nuclear response starts with the all-hazard building blocks but has scenario-specific nuances that require specialized capabilities and training. Significant new investments in resources will have to be made to develop and sustain adequate nuclear incident response and recovery capabilities at the local and state levels.

Highlights

- Renewed focus needs to be applied to preparing for nuclear incidents, domestically and internationally, to include updated protocols and technologies.
- A general lack of understanding of radiation and health aspects of nuclear and radiological events, especially when combined with misinformation, disinformation, and mal-information, will significantly complicate response and recovery efforts.
- There are continued shortfalls in local, state, and regional emergency management capabilities, particularly in trained and experienced personnel, for responding to and recovering from nuclear and radiological events.
- Given these limited capabilities, the private sector (including social media companies), nonprofit organizations, and civil society should be more deeply integrated into plans and exercises for responding to and recovering from nuclear and radiological incidents.

For a society to be resilient in these circumstances, government readiness should not be the only metric. A society needs an educated and informed citizenry that can prepare themselves in advance to be more resilient after such an attack. This is best guided by a concerted education campaign that presents a pragmatic explanation of the risks and helps the general population understand what they must do after an attack to safeguard themselves and their families. Domestic education programs for these two primary audiences (FSLTT officials and the general population) must be complementary to maximize the nation's readiness and resilience. Further, it is important to consider pre- and post-attack education as distinct topics.

> FINDING 9-2: Updated protocols and technologies, including forensics, are needed to support pre-event warning and to respond to an announced or threatened use of a nuclear device. Current protocols and technologies are not keeping up with emerging technologies.

The response capabilities following a pre-event warning of a nuclear attack on the U.S. homeland, U.S. interests overseas, and treaty allies do not yet take advantage of various emerging technologies, including smart technologies enhanced by artificial intelligence (AI) and machine learning that could be harnessed to upgrade the existing indicators and warning systems (PANO 2022).

Countermeasures include diagnostics/forensics to support attribution in the event of the threat of a nuclear attack (e.g., blackmail) or in the aftermath of an attack. A capable and "advertised" nuclear forensics program can play a deterrent role. The federal post-detonation nuclear forensics program collects data and samples and contributes to weapon or device attribution. Recently, a National Academies of Sciences committee reviewed the forensics program and made a number of recommendations to strengthen it (National Academies of Sciences 2021). This committee concurs with those recommendations.

Preparing to respond to terrorists' announced/threatened use of a nuclear weapon or device has been a significant component of the National Nuclear Security Administration (NNSA) Nuclear Emergency Support Team (NEST)'s Joint Technical Operations Team (JTOT) program since its inception (Tilden and Boyd 2023). NNSA has developed a variety of tools and techniques to search for, diagnose, assess, defeat, transport, dismantle, and attribute nuclear devices while retaining chain-of-custody control to facilitate subsequent legal action. The formal pre-detonation forensics program contributes to weapon attribution facilitated by intelligence-informed databases of weapon characteristics. The NNSA laboratories have expanded their efforts to understand how an improvised nuclear device (IND) could be designed using nontraditional materials in order to develop appropriate countermeasures.

In the event of the actual employment/use of a weapon, the United States maintains a capability to respond to a nuclear weapon or RDD detonation. These capabilities include NNSA's NEST and programs developed by the Department of Defense, FEMA, and the Centers for Disease Control and Prevention (CDC). Efforts have been made to establish a whole-of-government (i.e., federal, state, local) approach to addressing public health concerns and providing population movement recommendations in the detonation zone and adjacent areas.

9.2 MESSAGING DURING NUCLEAR INCIDENT

FINDING 9-3: A general lack of understanding of radiation and health aspects of nuclear and radiological events—especially when combined with misinformation, disinformation, and mal-information (MDM)—could significantly complicate response and recovery efforts. Maintaining a sufficient number of well-trained, trusted nuclear experts at the state and local levels will be essential to manage public communications during the response and recovery efforts of an impacted community.

There are significant challenges to informing and instructing the impacted and wider populace in the immediate aftermath of a nuclear event. Efforts to provide warning and lifesaving information for a nuclear incident have not kept pace with emergency information systems that have been developed for other disasters. For example, the National Oceanographic Atmospheric Administration maintains separate warning systems for tsunamis, tornados, and hurricanes, and the U.S. Geological Survey is developing an embryonic early warning system for earthquakes. Natural disasters happen with such frequency that they are the primary focus of exercises and allow crisis managers to gain familiarity with what they should do when these disasters occur. This is not the case with nuclear/radiological events where the absence of a previous attack, and its low probability, makes preparing, exercising, and communicating in ways that will save lives and facilitate recovery very difficult.

In the event of a nuclear or radiological attack, government officials and the public are likely to be unfamiliar with where they can go to get reliable information to include even the most rudimentary guidance such as whether to shelter in place or to evacuate. Those living near an attack site are likely to have severely disrupted communications capabilities, adding another challenge to information dissemination. This is fertile ground for the kind of harmful consequences associated with MDM that could be spread over social media, as was evident during the COVID-19 pandemic. Countering any MDM will take a substantial investment to develop a prophylactic "counter disinformation" capability to ensure a coordinated response that nimbly provides authoritative/trusted information. During the Fukushima Daiichi accident and response, the spokespeople had backgrounds in food, epidemiology, health and medicine, and environmental expertise. They provided critical technical information to decision-makers, responders, and the impacted communities alike. Such programs must be established, maintained, and exercised prior to a nuclear incident to avoid or minimize the MDM challenge. Countering disinformation will require utilizing all mechanisms (radio, TV, social media) to communicate authoritative/trusted messages and establish techniques to counter MDM.

Radiological operations support specialists (ROSS) are experts in radiological and emergency response who provide technical expertise to state, local, tribal, and territorial (SLTT) jurisdictions (Irwin 2022). This FEMA-led program, first piloted in nuclear

BOX 9-2
Immediate Messaging

Joseph Pfeifer and James Schwartz were first responders on 9/11. Chief Pfeifer was the first fire battalion chief to arrive at the World Trade Center. Responding with members of Engine 7 and Ladder 1, he implemented a command center in the North Tower's lobby. Under Chief Schwartz, the Arlington County Fire Department led the Unified Command effort to the attack at the Pentagon.

When discussing the necessary and immediate messaging in response to a nuclear incident, they gave the committee this simple and meaningful advice.[a]

"In an active shooter incident we tell the public to run, hide, fight. Well, what are we going to tell the public to know for an IND (improvised nuclear device)?

And what we thought about was again 3 words, and I throw this out [...] to all of you [...] well, what about,

Hide?

Help?

Go?

The first is hide, to go inside. Help each other, because the first responders aren't coming, at least initially. And then, when directed, go, move out of the area.

So Hide, Help, Go."

[a] Written materials submitted to a study committee by external sources and public meeting recordings are listed in the project's Public Access File and can be made available to the public upon request. Contact the Public Access Records Office (PARO) at the National Academies of Sciences, Engineering, and Medicine for a copy of the list and to obtain copies of the materials. E-mail: paro@nas.edu.

detonation exercises in 2014, gathers and interprets data to assist decision-making and provides information about available federal assets to the SLTT first responders, key leaders, and decision-makers. FEMA provides public affairs and National Incident Management System training to these nuclear subject matter experts (SMEs) so that they can operate effectively during an emergency. Maintaining a sufficient number of well-trained nuclear experts who can be readily available for SLTT organizations to draw upon will be key to getting timely and accurate information directly to an impacted community.

9.3 STATE, LOCAL, TRIBAL, AND TERRITORIAL CAPABILITIES AND NEEDS

FINDING 9-4: There are continued shortfalls in SLTT and regional emergency management capabilities for responding to and recovering from nuclear and radiological events. Given these limited capabilities, state and local authorities would benefit from more closely integrated efforts by the various federal departments and agencies, as well as a central repository to provide support before, during, and after a nuclear incident. Achieving this integration will require strong leadership at the White House level.

Current and past strategies for managing a nuclear incident have focused almost exclusively on the immediate alert and preparing for just the first minutes of response. As this is clearly insufficient, efforts are underway to develop response capacity for at least the first 72 hours following an event. The updated Nuclear Detonation Planning Guide (Federal Emergency Management Agency 2022) provides 72-hour guidelines (Federal Emergency Management Agency 2023). There are also 100-Minute Guidelines (U.S. Department of Homeland Security 2021) and National Council on Radiation Protection and Measurements guidelines for volunteers and non-radiation workers (National Council on Radiation Protection and Measurements 2022a[1]; National Council on Radiation Protection and Measurements 2022b[2]). Additionally, there are training videos that provide public and responder education (Centers for Disease Control and Prevention 2023[3]) and a variety of public health training courses developed by the

[1] The National Council on Radiation Protection and Measurements (NCRP) has recently issued two Statements addressing volunteers and non-radiation workers that have not been previously covered in other NCRP reports. NCRP Statement #14 provides recommendations for maintaining the readiness of radiation detection equipment retained by municipal, county, and state entities, including fire services, law enforcement, emergency management, public health agencies, and hospitals. A three-tiered, mission-oriented approach is described, which allows users to attain confidence in their equipment while working within available funding and personnel resources. It recognizes that a functional instrument, even if not formally calibrated, can still support certain missions during a large-scale emergency and is preferred to an absence of instrumentation.

[2] NCRP Statement #15 recommends a tiered approach for respiratory protection for a subgroup of emergency workers, including public health and mass care workers and volunteers, who may be at risk of an inhalation or incidental ingestion hazard generated by arrival and movement of potentially contaminated people. Workers and volunteers will be involved in assisting the displaced population with evacuation, screening and decontamination at community reception centers (CRCs), mass care at public shelters, and referrals for medical, relocation, and other services outside the affected area. Respiratory protection standards and guidelines that are suitable for occupational exposure scenarios will be difficult to implement for all workers who will be interacting with and providing services to potentially contaminated people. Unlike the medical response to mass casualties or crises, there is currently no equivalent provision for "crisis standards of care" when it comes to addressing the health and safety needs of this group of emergency workers. Lack of guidance and potential confusion about acceptable approaches to protect their health and safety can impede emergency response operations.

[3] An example of this is the efforts CDC has put into creating animated videos to address (1) shelter in place during a radiation emergency that highlights how shelter in place would be different for a radiation emergency compared with shelter in place instructions during COVID, (2) exposure versus contamination in the radiation world that explains the difference between exposure and contamination and highlights the differences with COVID, and (3) understanding potassium iodide, clearing some of the misconceptions about its use (specifically those arising from the conflict in Ukraine). Animated videos can be found at https://www.cdc.gov/radiation-emergencies/communication-resources/educational-videos.html

Centers for Disease Control and Prevention (Centers for Disease Control and Prevention 2022)[4] that are available to the SLTT response community.

These training efforts are commendable but would be more useful if they were incorporated into a central repository where the full range of information needed by SLTT responders and authorities could be accessed. A central repository would have to be managed initially at the White House level by a designated individual or interagency group but should eventually be turned over to a lead federal agency such as FEMA or the CDC. The U.S. Department of Health and Human Services maintains Radiation Emergency Medical Management (REMM), but REMM does not have an overall strategy for creation of information and capabilities. A commitment to long-term funding will be needed to train responders/receivers in these integrated protocols. In addition, there needs to be a sustained, focused examination of roles and responsibilities connected with transitioning from response to recovery. This has yet to be done in any systematic way. The United States has been through enough large-scale catastrophes to develop an understanding of what is involved in crisis response and what is required to have a successful post-disaster recovery. Capturing this knowledge would provide the basis to update national strategies for radiological or nuclear emergency response and recovery.

Following an attack, prompt post-event actions are necessary to communicate risk information, save lives, support forensics, and guide recovery. These actions include:

- **Partnerships.** A response to a nuclear or radiological emergency places a premium on effective coordination and collaboration among local and state authorities and federal and regional-based partners.[5] National planning scenarios and the development of matrices and annexes that outline national chemical, biological, radiological, and nuclear (CBRN) response capabilities are foundational to identifying the roles and responsibilities of each partner. Enhanced preparedness requires moving from planning scenarios to conducting major exercises and analyzing the outcomes to identify any gaps in capabilities, tools, training, and organization. The actual decision-makers for a real event—that is, governors, county commissioners, mayors and other state and local officials with delegated authorities—should participate in these exercises. This will provide a basis for determining which entities are best suited to perform which critical functions within a state, county, and municipality. Importantly, managing a nuclear emergency should draw on extant capabilities that are used in more commonly occurring disasters such as floods, wildfires, and hurricanes. Additionally, the knowledge derived from radiological and nuclear

[4] The CDC is piloting a new training course titled "Applied Course for Public Health Decision Making in a Radiation Emergency" specifically designed for the public health community that uses a holistic approach to emergency preparedness and response.

[5] These partners can include the Federal Bureau of Investigation (FBI) Stabilization Teams, the FBI WMD coordinator, Department of Homeland Security field partners, Department of Energy (DOE) Radiological Assistance Program teams, DOE national labs, state Radiation Safety Offices, state and regional HAZMAT teams, state-level law enforcement, and others.

preparedness planning, particularly with respect to informing regional and national collaborative efforts, can enhance integrated response capabilities and recovery efforts for major natural disasters. Establishing ongoing and productive partnerships centered on an all-hazard approach is a crucial element in achieving effective preparedness.

- **Procurement and Supply Chain.** Maintaining large inventories of nuclear response equipment (e.g., monitoring equipment, personal protective equipment [PPE], medical supplies, and dosimeters) in every community is not practical. As such, repositories of these resources maintained at the national and regional levels is more practical but places a premium on rapid distribution to impacted areas when they are needed. Supply chain management would be an important element of establishing and managing such repositories. As the COVID-19 pandemic highlighted, many supply chains depend on foreign or single-source suppliers and are susceptible to disruption. Supply chain management to include the identification of vulnerabilities and implementing approaches to addressing them (e.g., preserve PPE, accelerate distribution, and pivot to allocation) is critical to ensuring an effective nuclear incident response.[6] The relevant government agencies that procure the equipment must make their requirements clear and provide for emergency suspension or waiver of any rules or statutes, such as the Jones Act. Risk matrixes to help decision-makers deal with supply chain disruptions, without generating unintended consequences, will be needed as will effective information-sharing protocols.
- **New medical diagnostics and countermeasures.** The combination of innovative medical diagnostics and treatments for radiation exposure can save many lives that would be at risk even when excellent preparedness efforts lead to timely rescue from an area impacted by a nuclear detonation. Given the large number of casualties, an integrated clinical diagnostics system to enhance surge capacity is needed. One example would be to apply basic hematology techniques and lymphocyte depletion kinetics alongside dicentric assessments and other novel dosimetry methods in a tiered triage approach so that limited resources and time-consuming analysis can be focused on those in most dire need of treatment. Similarly, a combination of existing medical countermeasures and advanced treatments such as nucleic acid amplification and cell therapy for radiation injuries should be used in a tiered approach given the potential for large number of patients, the time-sensitive nature of the treatment, and the limited number of trained medical providers (Coleman et al. 2013). These diagnostic and therapeutic elements should serve as the foundation of a national concept of operations

[6] A negative element within every response is the subset of citizens who cash in on the vulnerability of the public during a disaster. Updated strategies should address the need to combat various forms of profiteering when there is a sudden surge in demand and the legitimate supply chain cannot quickly ramp up. The public is particularly vulnerable to a shortage of needed supplies, such as personal protective equipment, that can be misrepresented or counterfeited. During a crisis, monitoring and prosecuting such cases becomes difficult to prioritize.

(CONOPs) to integrate and sustain capabilities across the government and private sector (U.S. Department of Health and Human Services 2023).

9.4 COMMUNICATIONS AFTER A NUCLEAR INCIDENT

FINDING 9-5: Existing approaches for managing and disseminating trusted, accurate, science-based information in response to nuclear and radiological incidents are not adequate. The erosion in public trust in government could hamper the effective communication of public health and safety guidance in times of crisis.

There are nascent steps in considering social media and other alternative information pathways when advising the public on what to do during a nuclear or radiological incident, but these are not as engrained in the public's mind as long-standing approaches. Moreover, weapons of mass destruction (WMD) policies and operational planning for the federal government date back many decades and are exercised with state and local counterparts only in a limited way. These policies and plans were not designed to counter MDM, which is endemic in today's environment.

Overall, the individual citizen has not factored into response planning since civil defense measures were abandoned in the mid-twentieth century. In the aftermath of a nuclear incident, however, individual citizens must determine: (1) whether they shelter in place or evacuate, and by what route, when they receive warning of an imminent attack; (2) which sources of information they will listen to for guidance; and (3) how comfortable they are about returning to an area that they had to leave because of radiation exposure. Waiting until there is an incident to provide lifesaving information to the public is a prescription for disaster. The appropriate planning approach is to embrace the tenet that only an "informed public" will be in a position to make sound decisions following a nuclear incident. Preparedness efforts should place an emphasis on public awareness and education efforts. This will require taking response plans off the shelf and getting them into the hands of local decision-makers and the general public. One model for such an approach is to replicate on a much wider scale the kind of guidance, testing, and exercises that are done with residents and local officials who are located in the vicinity of nuclear power plants.

Communicating public health threats and harm-mitigation strategies must take into account cultural, language, and technical barriers that may interfere with the ability of vulnerable populations to understand and act on that information. For instance, these populations may have limited access to broadband, which limits the reach of digital-based sources of emergency preparedness, response, and recovery information. Analyzing and addressing the barriers for diverse populations to comprehend and effectively use information is a prerequisite to any sound community engagement effort for advancing nuclear and radiological preparedness.

In the absence of readily available federal guidance, states and local leaders will make their own decisions. This underscores the need for a flexible response framework that enables states to make wise independent decisions while working alongside federal planning and response efforts. Since the State Radiation Safety Control Program is typically responsible for managing radiological emergencies in most states, it is essential to integrate these offices into federal planning activities. The State Radiation Control Program operates under the umbrella organization known as the Conference of Radiation Control Program Directors (CRCPD) (Conference of Radiation Control Program Directors 2023), which serves as a convenient centralized resource to access state programs and their experts. While the Environmental Protection Agency (EPA), CDC, and Department of Energy (DOE) routinely provide funding for various advisory and technical initiatives, the current collaboration with CRCPD, often ad hoc, relies on relationships rather than a programmatic structure.

9.5 RECOVERY FROM A NUCLEAR INCIDENT

> FINDING 9-6: Medium- and long-term plans for any impacted areas are inadequate to recover from nuclear and radiological attacks on the U.S. homeland and abroad. Prior to a crisis, nuclear and radiological hazards must be understood to prioritize response and recovery actions, particularly those requiring time-sensitive decisions. To date, there is no consensus among federal, state, and local authorities on standards for radiation cleanup and area rehabilitation, possibly impeding a "return to normal" for an affected community.

Beyond preparing for the immediate response, there must also be plans and capabilities to manage the transition from response to long-term recovery to contain and restore an impacted area after a nuclear or radiological attack. Prior to a crisis, state and county officials and responders must understand nuclear and radiological hazards to prioritize response and recovery actions, particularly those requiring time-sensitive decisions. The goal should be to ensure that local and state leaders have a sound framework for decision-making. Disadvantaged populations need particular attention as the vulnerabilities and the need for external support will be greater. An effective whole-of-community approach to preparedness should include the trusted interlocutors, such as community nonprofit organizations and faith leaders, for marginalized groups.

Internationally, while the United States consistently offers its assistance, such as during the Fukushima incident, there are not adequate plans to guide how an impacted state or municipality could accept and utilize assistance from international partners. Part of the challenge is the lack of knowledge regarding the specific nature of the assistance and the difficulty of planning for its integration into local and state responses. The International Atomic Energy Agency (IAEA) Response and Assistance Network (RANET) is a well-organized resource that could help address this significant challenge.

There are plans and capabilities in place to support a U.S. response to a nuclear incident overseas. This includes the International CBRN Response Protocol that spells out the roles and responsibilities of the relevant federal departments and agencies for nuclear or radiological response overseas. Long-term support to allies for recovery will be managed through the established mechanisms within the relevant federal departments and agencies, particularly the U.S. Departments of State and Commerce.

Within the national borders of the United States, there have been many lessons learned from the February 2023 East Palestine, Ohio, rail disaster; the COVID-19 pandemic; civil unrest; and climate-related catastrophes of mega wildfires, hurricanes, floods, and tornadoes. While Americans are typically good at managing the life and safety issues associated with disaster, recovery efforts are too often developed on the fly and, as a result, are poorly conceived and executed. Plans and protocols to transition from disaster response to recovery should be improved or developed where they do not exist. For instance, there are often plans in place for evacuating neighborhoods or a region but no formal planning for returning residents to impacted areas. The transition from an emergency to "normal" conditions requires considerable time and effort and needs ongoing public communications. A nuclear or radiological incident will involve particularly challenging and largely unsettled issues associated with post-event cleanup. Today there is still no consensus among federal, state, and local authorities on standards for radiation cleanup and area rehabilitation. This will likely serve as a major impediment to a "return to normal" for areas affected by contamination.

9.6 AN ALL-OF-SOCIETY APPROACH

> FINDING 9-7: Significant roles exist for the private sector (including social media companies), nonprofit and faith-based organizations, and civil society, in responding to and recovering from nuclear and radiological incidents. All have incentives to speed up the recovery process by providing assistance.

The value that volunteers from the community, the surrounding region, and the country can provide should not be underestimated. For a major disaster, there are never enough available professionals to meet all the needs of the impacted populations. Plans that anticipate and embrace volunteers throughout the disaster cycle are instrumental for successful response and recovery. Operational planning needs to account for the role of volunteers and factor in what assistance they can provide. Community associations, faith-based organizations, and other local volunteer organizations should be part of the planning process. Additionally, representatives from these groups should be invited to participate in exercises and professional conferences to bolster an understanding of the role and capabilities they can bring to bear. Nongovernmental organizations and private sector leaders should be consulted during planning and exercises and throughout a nuclear/radiological emergency. The National Alliance for Radiation Readiness (NARR) is an important organization representing practitioners in the field of radiation readiness, including state and local public health practitioners, elected officials at the state and local level, and first-responder/first-receiver groups.

Inevitably, there will not be enough highly skilled and equipped teams like DOE's Radiological Assistance Program (RAP) team, EPA's Radiological Emergency Response Team (RERT), and the National Guard Weapons of Mass Destruction Civil Support Teams. Thus, plans that incorporate volunteers are essential for a successful response and recovery.

RECOMMENDATION 9-1: FEMA should reinvigorate a dynamic, comprehensive, and inclusive exercise regimen, in coordination with the Federal Radiological Preparedness Coordinating Committee (FRPCC) and with guidance and oversight from the NSM-19-established council of leadership. This should include fully utilizing the FRPCC (Federal Emergency Management Agency 1996; Nuclear Regulatory Commission 1973) in its capacity as a national-level forum to develop and coordinate radiological prevention and preparedness policies and procedures.

Planners need to insert unscheduled crises into the normal exercise cycle that challenge senior decision-makers to grapple with multiple emergencies at once while responding to a radiological emergency. Training protocols can be designed to assist practitioners in dealing with dynamic situations where there is considerable uncertainty about outcomes. Exercises can include identifying gaps in available capabilities for which decision-makers will need to develop potential workarounds. Businesses with advanced supply chain capabilities could provide helpful insights and should be included in exercises and training.

Online tools and local websites have been developed by many state emergency management agencies and regional preventive radiation and nuclear detection focus groups. These tools and websites should be used and assessed during exercises along with training that incorporates lessons learned to improve crisis response. Each exercise is also an opportunity to educate the participating leaders and the local community and should include a strong public affairs component. By thoughtfully engaging traditional and social media, this public affairs outreach can provide subject matter familiarization and important information and guidance to local community members that will be lifesaving during an actual event. It can also provide reassurance to the public that their elected leaders are engaged along with pre-identified and well-trained local subject matter experts who can build trust with the public on radiation-related issues.

RECOMMENDATION 9-2: FEMA with CDC, EPA, DOE, and NIH should empower local response, by making available simple and accessible real-time information through application development that will facilitate standardized actions and guide an appropriate public response. To assist, the White House should clarify the agency that serves as overall lead for providing federal interagency coordination and oversight of developing response tools to include educating state, local, tribal, and territorial officials as well as the general public on their availability and utility, and it should strive to establish itself as a trusted agent.

User-friendly data visualization products such as graphs, maps, and infographics will be essential tools in a radiological emergency. As with many low-probability/high-consequence events, there is no single platform or network to connect the non-governmental radiological/nuclear community to media or public officials. It will be important to develop such a mechanism to share resources and tools, including technical methods and information.

RECOMMENDATION 9-3: The president should request and Congress should support adequate resources for consequence management (CoM) programs that are key to a nuclear incident response. This should recognize the important role states, localities, tribal nations, and territories play in saving lives. More resources are needed because these programs have insufficient budgets, staffs, and capabilities, and yet are foundational to any successful response to a nuclear or radiological event.

References

Bentz, Julie. 2022. "Learning from Catastrophe: Lessons from the COVID-19 Pandemic for Preparing for and Responding to a Domestic Radiation Emergency." *NTI Paper*. https://www.nti.org/analysis/articles/learning-from-catastrophe-lessons-from-the-covid-19-pandemic-for-preparing-for-and-responding-to-a-domestic-radiation-emergency/.

Centers for Disease Control and Prevention. 2022. "Radiation Emergency Training, Education, and Tools." Last Modified October 20, 2022. https://www.cdc.gov/radiation-emergencies/php/training/?CDC_AAref_Val=https://www.cdc.gov/nceh/radiation/emergencies/training.htm.

Coleman, C. N., D. J. Blumenthal, C. A. Casto, M. Alfant, S. L. Simon, A. L. Remick, H. J. Gepford, T. Bowman, J. L. Telfer, P. M. Blumenthal, and M. A. Noska. 2013. "Recovery and Resilience after a Nuclear Power Plant Disaster: A Medical Decision Model for Managing an Effective, Timely, and Balanced Response." *Disaster Medicine and Public Health Preparedness* 7 (2): 136–145. https://doi.org/10.1017/dmp.2013.5.

Conference of Radiation Control Program Directors. 2023. "CRCPD A Partnership Dedicated to Radiation Protection." https://www.crcpd.org/.

Federal Emergency Management Agency. 1996. "Emergency Management and Assistance." Washington: U.S. Government Publishing Office. https://www.govinfo.gov/content/pkg/CFR-1996-title44-vol1/html/CFR-1996-title44-vol1-part351-subpartB.htm

---. 2022. "Planning Guidance for Response to a Nuclear Detonation," Third edition. Edited by Biological Chemical, Radiological, and Nuclear (CBRN) Office. https://www.fema.gov/sites/default/files/documents/fema_nuc-detonation-planning-guide.pdf

---. 2023. "Nuclear Detonation Response Guidance: Planning for the First 72 Hours." https://www.fema.gov/sites/default/files/documents/fema_cbrn-72-hournuclear-detonation-response-guidance.pdf

Irwin, William. 2022. "The Radiological Operations Support Specialist (ROSS)." National Radiological Emergency Preparedness Conference, Nashville, Tennessee, April 11, 2022. http://www.nationalrep.org/2022Presentations/Session1_ROSS_Presentation.Final.pdf.

National Academies of Sciences, Engineering, and Medicine. 2021. *Restoring and Improving Nuclear Forensics to Support Attribution and Deterrence: Public Summary*. Washington, DC: The National Academies Press.

National Council on Radiation Protection and Measurements. 2022a. *NCRP Statement No. 14–Instrument Response Verification and Calibration for Use in Radiation Emergencies*. National Council on Radiation Protection and Measurements (National Council on Radiation Protection and Measurements). https://ncrponline.org/wp-content/themes/ncrp/PDFs/2022/Statement_14.pdf.

---. 2022b. *NCRP Statement No. 15–Respiratory Protection Recommendations for Workers and Volunteers Responding to a Nuclear Incident Outside the Affected Area*. National Council on Radiation Protection and Measurements (National Council on Radiation Protection and Measurements). https://ncrponline.org/wp-content/themes/ncrp/PDFs/2022/Statement_15.pdf.

Nuclear Regulatory Commission. 1973. "10 CFR Part 73—Physical Protection of Plants and Materials." https://www.nrc.gov/reading-rm/doc-collections/cfr/part073/index.html

PANO. 2022. "Introducing Pano Actionable Intelligence for Wildfire Management." https://www.pano.ai/.

Tilden, Jay A., and Dallas Boyd. 2023. "Nerds, Ninjas, and Neutrons: The Story of the Nuclear Emergency Support Team." *Bulletin of the Atomic Scientists* 79 (2): 87–94. https://doi.org/10.1080/00963402.2023.2178179.

U.S. Centers for Disease Control and Prevention. 2023. "Radiation Emergencies–Resource Library." https://www.cdc.gov/nceh/radiation/emergencies/resourcelibrary/videos.htm.

U.S. Department of Health and Human Services. 2023. "MedicalCountermeasures.gov CBRN Radiological and Nuclear Medical Program." https://medicalcountermeasures.gov/barda/cbrn/radiological-and-nuclear-countermeasures/.

U.S. Department of Homeland Security. 2021. "Radiological Dispersal Device (RDD) Response Guidance Planning for the First 100 Minutes Report and Video." Last Modified January 12, 2023. https://www.dhs.gov/publication/st-frg-rdd-response-guidance-planning-first-100-minutes.

Appendix A

Presentations at the Committee's Public Information-Gathering Sessions

COMMITTEE ON ASSESSING AND IMPROVING STRATEGIES FOR PREVENTING, COUNTERING, AND RESPONDING TO WEAPONS OF MASS DESTRUCTION TERRORISM: NUCLEAR THREATS

Information Gathering Meeting #1
JUNE 21, 2022

- **DOE National Nuclear Security Administration Counterterrorism and Counterproliferation (NA-80)**
  Dallas Boyd
- **DOE National Nuclear Security Administration I Defense Nuclear Nonproliferation (NA-20)**
  Christine Bent
- **Overview of the MacArthur Foundation Project on Bolstering Counter-Proliferation Efforts within the Global Supply System**
  (co-principal investigators John Holmes and Stephen Flynn) Prof. Stephen Flynn (for Captain John Holmes, U.S. Coast Guard [ret.])
- **Nuclear Threat Initiative, Senior Director Nickolas Roth**
  Nickolas Roth, senior director, Nuclear Threat Initiative's Nuclear Materials Security Program Team
- **Belfer Center for Science and International Affairs (Harvard Kennedy School), program director Simon Saradzhyan**
  Simon Saradzhyan, program director, Belfer Center, U.S.-Russia Initiative to Prevent Nuclear Terrorism
- **Boston University Frederick S. Pardee School of Global Studies, Prof. Jessica Stern**
  Jessica Stern's Homeland Security Experts Group and Fellow Quincy Institute for Responsible Statecraft.

COMMITTEE ON ASSESSING AND IMPROVING STRATEGIES FOR PREVENTING, COUNTERING, AND RESPONDING TO WEAPONS OF MASS DESTRUCTION TERRORISM: NUCLEAR THREATS

Information Gathering Meeting #2
AUGUST 15, 2022

- **The Changing Nuclear Eras**
  Jane Lute, president and CEO, Council on Cybersecurity strategic director SICPA North America
- **The Role of NGOs to Counter Nuclear Terrorism**
  Emma Belcher, president of Ploughshares Fund
- **Roundtable Discussion**
  Uri Friedman Atlantic Council, *The Atlantic*
- **Roundtable Discussion**
  David Sanger, *The New York Times*

COMMITTEE ON ASSESSING AND IMPROVING STRATEGIES FOR PREVENTING, COUNTERING, AND RESPONDING TO WEAPONS OF MASS DESTRUCTION TERRORISM: NUCLEAR THREATS

Information Gathering Meeting #3
NOVEMBER 16, 2022

- **Worker And Public Health and Safety Risks**
  Kathryn Higley, Oregon State University
- **Advanced Nuclear Reactor Safeguards and Security**
  Ken Luongo, Partnership for Global Security
  Per Peterson, Kairos Power and University of California
  Jack Edlow, Edlow International Company
- **U.S. Foreign Obligations on Exported Nuclear Material, Nuclear Regulatory Commission Panel**
  James Rubenstone, chief, Material Control and Accounting Branch, Office of Nuclear Material Safety and Safeguards (NMSS)
  Barry Miller, senior international policy analyst, Office of International Programs (OIP) and Cynthia Jones, senior technical advisor for nuclear security, Office of Nuclear Security and Incident Response (NSIR)
  Rebecca Richardson, deputy director, Division of Physical and Cyber Security Policy, NSIR
  Desiree Davis, acting chief, Materials Security Branch, NSIR

COMMITTEE ON ASSESSING AND IMPROVING STRATEGIES FOR PREVENTING, COUNTERING, AND RESPONDING TO WEAPONS OF MASS DESTRUCTION TERRORISM: NUCLEAR THREATS

Information Gathering Meeting #4
NOVEMBER 29, 2022

- **Opening Remarks by DNN**
  Kasia Mendelson
- **Global Material Security (GMS)**
  Art Atkins, Daniel Abeyta, Christine Bent, Allison Johnston, and Kristin Hirsch, International Nuclear Security, Radiological Security Nuclear Smuggling Detection and Deterrence
- **Materials Management and Minimization (M3)**
  Jeffrey Chamberlin, Joan Dix, and Tiffany Blanchard–Case Nuclear Materials Removals Research Reactor Conversions

COMMITTEE ON ASSESSING AND IMPROVING STRATEGIES FOR PREVENTING, COUNTERING, AND RESPONDING TO WEAPONS OF MASS DESTRUCTION TERRORISM: NUCLEAR THREATS

Information Gathering Meeting #5
DECEMBER 6, 2022

- **9/11 First Responder, World Trade Center**
  Joseph Pfeifer, retired assistant chief of the New York City Fire Department, chief of counterterrorism and emergency preparedness (retired)
  Senior fellow with the Program on Crisis Leadership at Harvard Kennedy
- **9/11 First Responder, The US Pentagon**
  James Schwartz, retired chief of Arlington County Fire Department and Deputy County Manager, Arlington County, senior fellow with the Program on Crisis Leadership at Harvard Kennedy
- **Response and Recovery to Nuclear Terrorism**
  Orly Amir, Department of Homeland Security, program manager, National Urban Security Technology Laboratory
  Adela Salame-Alfie and Armin Ansari, Centers for Disease Control and Prevention, Radiation Studies Section
- **Response and Recovery to Nuclear Terrorism Federal Emergency Management Agency (FEMA)**
  Jessica Wieder, deputy director, Communications Division
  Jonathan Gill, physical scientist
  Joselito Ignacio, acting director and public health advisor
  Janis McCarroll, senior public health advisor

COMMITTEE ON ASSESSING AND IMPROVING STRATEGIES FOR PREVENTING, COUNTERING, AND RESPONDING TO WEAPONS OF MASS DESTRUCTION TERRORISM: NUCLEAR THREATS

Information Gathering Meeting #6
DECEMBER 15, 2022

- **Opening Address**
  Sara Cohen, deputy head of mission, Embassy of Canada to the United States
- **Experience across many time zones: Reflections of a Federal Responder**
  Daniel Blumenthal, attaché, U.S. Department of Energy
- **Assessing a Currently Changing Nuclear Landscape and the WMD Nuclear Terrorism Risk**
  Ambassador Kenneth Brill, founding director, U.S. National Counter-proliferation Center (2005–10), retired US Diplomat
  Brian Finlay, president and CEO, The Stimson Center
  Amy Woolf, nonresident senior fellow, Atlantic Council
  Moderator: Stephen Flynn, chair, National Academies WMD Nuclear Terrorism Committee
- **Bolstering International Efforts to Prevent Non-State Actors from Developing, Acquiring, Manufacturing, Possessing, Transporting, Transferring, or Using Nuclear Weapons and Their Means of Delivery**
  Nicki Mokhtari, United Nations Office of Preventing and Responding to WMD/CBRN Terrorism Unit
  Christian Carnus, criminal intelligence analyst, INTERPOL
  Richard Cupitt, The Stimson Center (retired)
  Moderator: Luke Hartig, President of National Journal Research, former senior director at the National Security Council
- **The Importance of International Cooperation**
  Ambassador Bonnie Denise Jenkins, under secretary for arms control and international security, U.S. Department of State
  Administrator Jill Hruby, under secretary of nuclear security and administrator, National Nuclear Security Administration (NNSA), U.S. Department of Energy
  Moderator: David Sanger, White House and national security correspondent, *The New York Times*
- **Resilience to a Nuclear Terrorism Incident**
  Kathleen Heppell-Masys, director general, Security and Safeguards, Canadian Nuclear Safety Commission
  Wendin Smith, director, Arms Control, Disarmament, WMD Non-Proliferation Centre, NATO
  Moderator: Michael Gresalfi, former senior advisor on counterproliferation and WMD threats to FEMA, DOE, and DHS
- **Roundtable Discussion with U.S. Allies, Panelists, and Committee Members**
  Moderator: Stephen Flynn, chair, National Academies WMD Nuclear Terrorism Committee

COMMITTEE ON ASSESSING AND IMPROVING STRATEGIES FOR PREVENTING, COUNTERING, AND RESPONDING TO WEAPONS OF MASS DESTRUCTION TERRORISM: NUCLEAR THREATS

Information Gathering Meeting #7
MARCH 6, 2023

- **Welcome**
  Drew Kuepper, strategy and analysis deputy assistant secretary, Department of Homeland Security
- **R/N Threat Assessment**
  Kevin Lehman, lead briefer, Department of Homeland Security
- **R/N Risk**
  Jennifer Pavlick, Management and Program Analysis, Department of Homeland Security
- **FY23–27 Global Nuclear Detection Architecture Strategy & DHS Inbound Nuclear Planning Framework**
  Major Greg Abide, strategic planner, Department of Homeland Security
  William Bilicic, strategy development branch chief, Department of Homeland Security
  Fredrick Breaux, Department of Homeland Security
- **DHS Component Capabilities Panel, DHS Components: CISA, CWMD, TSA, & USCG**
  Moderator, Theodore Macklin, president and founder, TOMAR Research, Inc.
- **CWMD Principal Deputy Assistant Remarks**
  Charles "Chas" Cook, principal deputy assistant secretary, Department of Homeland Security
- **ODNI Brief**
- **USSOCOM Brief**

COMMITTEE ON ASSESSING AND IMPROVING STRATEGIES FOR PREVENTING, COUNTERING, AND RESPONDING TO WEAPONS OF MASS DESTRUCTION TERRORISM: NUCLEAR THREATS

Information Gathering Meeting #8
MAY 2, 2023

- **Briefing**
  Melissa Dalton, assistant secretary of defense for homeland defense and hemispheric affairs, Department of Defense

Appendix B

Five Eras of Nuclear Terrorism[1]

Policymakers have thought about nuclear security since the early days of U.S. nuclear weapons development. While initially the emphasis was on protecting against espionage, discussions evolved to focus on other threats. Incidents like the terrorist attacks at the 1972 Munich Olympics, which demonstrated the danger of a small group of well-armed militants, focused international attention on the threat of non-state actors.[2] Soon after, the United States developed new approaches to protecting nuclear facilities from non-state actors, the International Atomic Energy Agency (IAEA) began developing physical protection recommendations, and the Convention on the Physical Protection of Nuclear Material (CPPNM)—which remains the only treaty that obligates countries to implement security measures for civilian nuclear material in international transport—was opened for adoption. It was not until the 1990s and the collapse of the Soviet Union, however, that governments began to consider preventing nuclear terrorism an international priority.

Since the early 1990s, there have arguably been five periods where high-level U.S. attention to international nuclear security has adapted and adjusted based on significant world events: the rapid response to the collapse of the Soviet Union, the post-9/11 era, the nuclear security summit era, the post-summit era, and the era of rapidly evolving risks.

---

[1] In response to the committee's mandate for this study found in the U.S. Congress William M. (Mac) Thornberry National Defense Authorization Act for Fiscal Year 2021, the committee felt it was important and valuable to stakeholders and decision-makers to put in context the historical eras for nuclear security. For this reason, the committee asked Nickolas Roth to summarize his presentation to the committee from June 21, 2022, as this appendix.

[2] Matthew Bunn, *Beyond Crises: The Unending Challenge of Controlling Nuclear Weapons and Materials*, February 27, 2012, https://npolicy.org/wp-content/uploads/2021/08/Beyond_Crises-The_Unending_Challenge_of_Controlling_Nuclear_Weapons_and_Materials.pdf.

B.1 RAPID RESPONSE

The first era was dominated by bilateral cooperation between two countries—the United States and Russia. Beginning in 1991, the United States led an emergency effort in response to the grave nuclear security risks emanating from the dissolution of the Soviet Union. Concerns about inadequate security and the risk of stolen nuclear weapons or materials amid the economic, political, and institutional crisis brought about by the Soviet collapse, led to the introduction of the Cooperative Threat Reduction (CTR) program, a multi-billion-dollar program of U.S. technical assistance to Russia and other former Soviet republics. Over its lifetime, CTR was enormously successful, leading to dramatic security improvements at Russian facilities with nuclear materials. In its early stages in particular, the signature characteristics of CTR projects were large financial investments, bilateral donor-recipient transactions. While ultimately unsustainable in either country, this work led to the development of strong personal relationships between Americans and Russians engaged in nuclear security.

B.2 9/11 RESPONSE

The second era of nuclear security began with the September 11, 2001, terrorist attacks that killed approximately 3,000 civilians in the United States. These unprecedented attacks demonstrated what organized, well-financed, sophisticated non-state actors, like Al Qaeda, could achieve. They also highlighted the dangers of complacency to low-probability, high-consequence threats. Motivated by evidence that Al Qaeda was pursuing nuclear weapons and had considered targeting U.S. nuclear power plants, the United States redoubled efforts to strengthen security at civilian and government nuclear facilities within its own borders.[3] The shock of the 9/11 attacks also motivated other countries to take action to strengthen their nuclear security.

During this period, the United States expanded its bilateral nuclear security cooperation with Russia. Beginning in 2005, under the Bratislava Nuclear Security Initiative, the United States and Russia intensified their collaboration focused on upgrading the security of Russian nuclear facilities, expanding emergency response, enhancing nuclear security culture, accelerating research reactor conversions and fuel repatriation, and sharing best practices.[4]

While the U.S.–Russian initiative still represented the bulk of the international nuclear security cooperation, there were further developments in bolstering international nuclear security architecture. In 2005, parties to the CPPNM agreed to an amendment that expanded the convention's scope to include the protection of nuclear materials

[3] National Commission on Terrorist Attacks Upon the United States, *The 9/11 Commission Report* (Washington, DC: National Commission on Terrorist Attacks Upon the United States, July 2004), https://www.9-11commission.gov/report/911Report.pdf.

[4] U.S. Department of Energy, *U.S. and Russia Complete Nuclear Security Upgrades under Bratislava Initiative* (Washington, DC: DOE, December 2008), https://www.energy.gov/articles/us-and-russia-complete-nuclear-security-upgrades-under-bratislava-initiative.

located in nuclear facilities dedicated to peaceful uses and to strengthen protection against sabotage of nuclear facilities. The amendment also requires a conference five years after entry into force to review implementation. The International Convention on the Suppression of Acts of Nuclear Terrorism was also negotiated and ratified in this period. Although dominated by the same bilateral dynamics as the first era of nuclear security, this second era expanded the focus to strengthening nuclear security in more countries.

B.3 NUCLEAR SECURITY SUMMITS

The third era began when the Obama administration made strengthening international nuclear security one of its signature policy priorities. Driven by the United States, the four nuclear security summits that took place from 2010 to 2016 shifted the focus of international nuclear security cooperation from the bilateral relationship between the United States and Russia to a much broader effort among dozens of nations and, most importantly, elevated nuclear security to the highest levels of governments around the world. Over this period, countries eliminated thousands of kilograms of nuclear materials, developed stronger design basis threats, adopted new insider threat protection measures, expanded performance assessment and testing, and adopted new security culture programs; and the international institutions and legal instruments bolstering international nuclear security were significantly strengthened.[5]

While this was occurring, however, U.S. cooperation with countries that possessed the most weapons-usable nuclear material and faced the most significant risks was either stagnant or in decline. In 2014, nuclear security cooperation at Russian nuclear facilities stopped after Russia invaded Ukraine.[6] The United States played an important role in creating China's nuclear security Center of Excellence, but that effort did not ultimately lead to expanded cooperation.[7] There was little progress in strengthening nuclear security cooperation with India and other countries with nuclear facilities or weapons-usable materials.

The Nuclear Security Summit era extended beyond the summits themselves. States followed through on commitments they made during the summit's months or, in some cases, years after the last summit in Spring 2016. For example, in September 2017, the National Nuclear Security Administration (NNSA) announced that it had removed the last batch of highly enriched uranium (HEU) from Kazakhstan's Institute of Nuclear

---

[5] For more on what was achieved, see Martin B. Malin, Matthew Bunn, Nickolas Roth, and William H. Tobey, "Advancing Nuclear Security: Evaluating Progress and Setting New Goals," *Belfer Center for Science and International Affairs,* March 18, 2015, https://www.belfercenter.org/publication/advancing-nuclear-security-evaluating-progress-and-setting-new-goals.

[6] See Matthew Bunn, "Steps for Rebuilding U.S.–Russian Nuclear Security Cooperation," *Belfer Center for Science and International Affairs,* July 2017, https://www.belfercenter.org/sites/default/files/files/publication/a434_1.pdf.

[7] The White House, *U.S.–China Joint Statement on Nuclear Security Cooperation* (Washington, DC: Office of the Press Secretary of the White House, March 2016), https://obamawhitehouse.archives.gov/the-press-office/2016/03/31/us-china-joint-statement-nuclear-security-cooperation.

Physics's VVR-K reactor; overall, more than 200 kilograms of HEU was removed from the facility.[8] In August 2017, South Africa converted its Mo-99 production from HEU to low-enriched uranium (LEU) targets and, in January 2018, The Netherlands converted its Mo-99 production to LEU targets.[9] Numerous countries continued to upgrade their physical protection requirements after the summits. Countries continued to join international agreements.[10]

B.4 POST-SUMMIT ERA

The fourth era of nuclear security began in 2018 after many of the commitments made during the nuclear security summit process had been accomplished. Building on the success of the summits proved elusive, and high-level attention to nuclear security waned. Despite some statements to the contrary, for the first time in over two decades, nuclear security did not appear to be an international priority. The U.S.-Russian bilateral cooperation that dominated the first years of nuclear security work had almost completely come to an end. The political momentum created by multilateral nuclear security summits subsided. The IAEA, which was supposed to be the focal point for international nuclear security efforts after the summit process, remains focused on nuclear energy, safety, and technology, and not as much on nuclear security. After a decade of freefall, budgets for U.S. nuclear security programs only showed modest growth primarily driven by Congress (see Figure B-1).

B.5 ERA OF RAPIDLY EVOLVING RISKS

As attention has declined to nuclear security, new challenges have emerged. The incorporation of digital technologies within nuclear facilities is creating new potential vulnerabilities, and emerging technologies like artificial intelligence and drones are providing adversaries with dangerous new capabilities. Increasingly destructive natural disasters like hurricanes and wildfires are generating new challenges for all aspects of nuclear security operations. Two recent crises, the COVID-19 pandemic and Russia's invasion of Ukraine, a country with a large nuclear infrastructure, have raised questions about how regulatory oversight, physical protection systems, and even international institutions should prepare for and respond to protracted anthropomorphic or naturally

[8] U.S. Department of Energy, *NNSA Partners with Kazakhstan Research Institute to Remove All of Its Highly Enriched Uranium* (Washington, DC: DOE, September 2017), https://www.energy.gov/nnsa/articles/nnsa-partners-kazakhstan-research-institute-remove-all-its-highly-enriched-uranium.

[9] U.S. Department of Energy, *NNSA's Molybdenum-99 Program: Establishing a Reliable Domestic Supply of Mo-99 Produced without Highly Enriched Uranium* (Washington, DC: DOE), https://www.energy.gov/nnsa/nnsas-molybdenum-99-program-establishing-reliable-domestic-supply-mo-99-produced-without.

[10] Sara Z. Kutchesfahani, Kelsey Davenport, and Erin Connolly, The Nuclear Security Summits: An Overview of State Actions to Curb Nuclear Terrorism 2010–2016 (Washington, DC: Arms Control Association and Fissile Materials Working Group, 2018), https://www.armscontrol.org/sites/default/files/files/Reports/NSS_Report2018_digital.pdf.

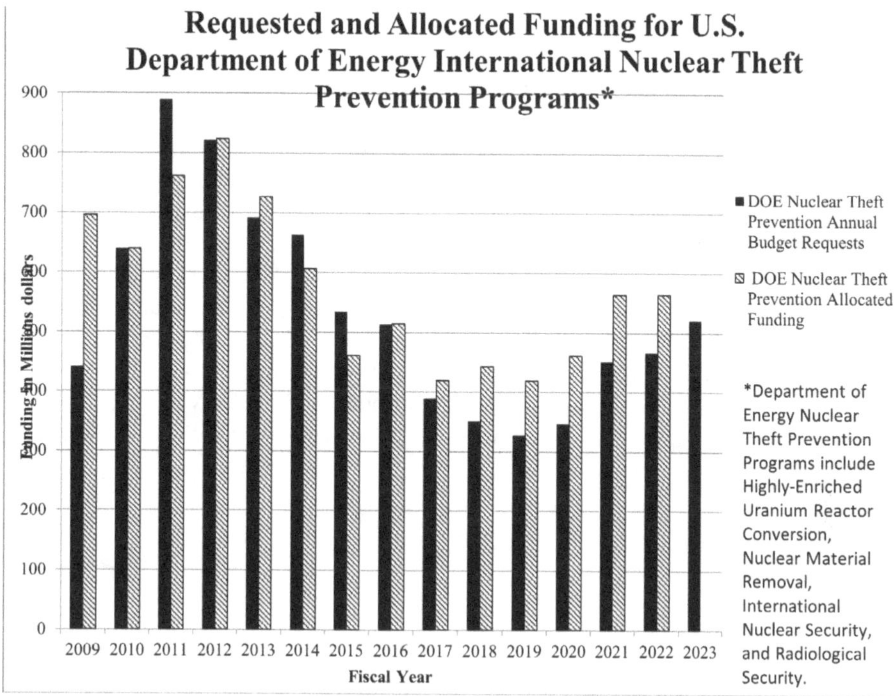

FIGURE B-1 Requested and allocated funding for U.S. Department of Energy International Nuclear Theft Prevention Programs.
SOURCE: Nickolas Roth, 2024.

occurring crises.[11] Domestically, political polarization is changing society in ways that challenge systems to detect and mitigate insider threats.

Beyond immediate challenges to physical protection systems, there are challenges to the political relationships that underpin the international nuclear security architecture. As the two countries with by far the most weapons-usable nuclear material, U.S. and Russian cooperation and leadership are essential to the viability of this architecture. Yet, Russia's invasion of Ukraine and occupation of Ukrainian nuclear facilities raises questions not only about Russia's commitment to nuclear security but also about the future of nuclear cooperation between the two countries. With relations between the United States and Russia at a historical nadir, it is unclear when, or if, nuclear security cooperation will resume or what form that cooperation would take in the future. Even areas of bilateral cooperation where norms are much stronger, like legally binding, verifiable limits on nuclear weapons deployments, are in doubt with Russia's decision to suspend its participation in the New Strategic Arms Reduction Treaty. Broader mul-

[11] Christopher Hobbs, Nickolas Roth, and Daniel Salisbury (2021), "Security under Strain? Protecting Nuclear Materials during the Coronavirus Pandemic," *RUSI Journal* 166: 2, 40–50, https://doi.org/10.1080/03071847.2021.1937302.

tilateral cooperation has also been impacted. The Global Initiative to Combat Nuclear Terrorism, which was co-chaired by Russia and the United States, has been suspended, leaving a gap in internationally coordinated nuclear security exercises.

Beyond the immediate implications of degraded bilateral relationships, there are important conceptual shifts in nuclear security. For decades, international nuclear security cooperation has been premised on the idea that non-state actors were the primary threat to nuclear facilities. While not the first incident where a state attacked a nuclear facility, Russia's invasion of Ukraine has changed that paradigm. In this new and evolving nuclear security model, governments will need to develop a novel policy toolkit for discouraging states from engaging in nuclear sabotage. Countries should be prepared for so-called beyond-the-design-basis threat scenarios. These are important not just for times of war but also in addressing the uncertainty of rapidly evolving technological risks or other political, economic, or natural disruptions. While COVID-19 and the war in Ukraine are the latest crises to impact nuclear facilities, they will certainly will not be the last.

Appendix C

Biographies of Committee, Consultant, and Study Director

Stephen E. Flynn (*Chair*) is a professor of political science and the founding director of the Global Resilience Institute at Northeastern University, where he leads a major university-wide transdisciplinary research initiative to inform and advance societal resilience in the face of growing human-made and naturally occurring turbulence. At Northeastern, he also holds affiliated faculty appointments in the College of Engineering and the School of Public Policy and Urban Affairs. Prior to September 11, 2001, Dr. Flynn served as an expert advisor to the U.S. Commission on National Security (Hart-Rudman Commission), and following the 9/11 attacks, he was the executive director of a blue-ribbon Council on Foreign Relations homeland security task force. He served as the principal advisor to the bipartisan Congressional Port Security Caucus, advised the Bush Administration on maritime and homeland security issues, and after the November 2008 election of President Barack Obama, served as the lead policy advisor on homeland security as a part of the presidential transition team. From 2003 to 2010 he served as a member of the National Research Council's Marine Board. Dr. Flynn has previously served as president of the Center for National Policy and spent a decade as a senior fellow for National Security Studies at the Council on Foreign Relations. Dr. Flynn was an active duty commissioned officer in the U.S. Coast Guard for 20 years, including two tours as commanding officer at sea. He is a coauthor of the textbook *Critical Infrastructures Resilience: Policy and Engineering Principles* (2018) and author of *The Edge of Disaster: Rebuilding a Resilient Nation* (Random House, 2007) and *America the Vulnerable* (HarperCollins, 2004). He has a presidential appointment to serve on the Board of Visitors at the U.S. Coast Guard Academy where he earned a bachelor of science in 1982. Dr. Flynn holds a master of arts in Law and Diplomacy and PhD degrees from the Fletcher School of Law and Diplomacy, Tufts University.

Madelyn R. Creedon (*Vice Chair*) is the president of Green Marble Group, LLC, a consulting company she founded after completing 36 years of federal service. She serves on a number of advisory and other boards related to national security, is the vice chair of the Secretary of Energy's Advisory Board, chair of the 2023 Strategic Posture Commission, a research professor at The George Washington University Elliott School of International Affairs and a nonresident senior fellow at the Brookings Institution. Prior to retirement, Creedon was principal deputy administrator of the National Nuclear Security Administration (NNSA) within the Department of Energy from 2014 to 2017. She served in the Pentagon as assistant secretary of defense for Global Strategic Affairs from 2011 to 2014, overseeing policy development in the areas of missile defense, nuclear security, cybersecurity, and space. She served as counsel for the U.S. Senate Committee on Armed Services for many years, beginning in 1990; assignments and focus areas included the Subcommittee on Strategic Forces as well as threat reduction and nuclear nonproliferation. During that time, she also served as deputy administrator for defense programs at the NNSA, associate deputy secretary of energy, and general counsel for the Defense Base Closure and Realignment Commission. She is currently a board member for the Nuclear and Radiation Studies Board at the National Academies of Sciences. Creedon holds a juris doctor from St. Louis University School of Law and a bachelor of arts from the University of Evansville.

Julie A. Bentz Member Julie Bentz retired in 2019 after a successful 33-year career, spanning active, reserve and National Guard commissioned service. She has been a recurring member of the White House National Security Council Staff and Homeland Security Council for the Executive Office of the President. While working at the White House, her roles included senior advisor for emerging technologies, deputy senior director for WMD, director of strategic capabilities, and director of nuclear defense policy. Major General Bentz currently serves on the Avista Energy Corporation Board of Directors. She also serves as chair of Sandia National Laboratory National Security Programs External Advisory Board and as a board member of Lawrence Livermore National Laboratory (LLNL) Global Security Directorate External Review Committee and LLNL's Weapons and Complex Integration Directorate External Review Committee. She is a member of the strategic advisory group for the CEO of CACI International, Inc. She also advises the Santiam Canyon Long Term Recovery Group, who assists those impacted by the September 2020 mega-fires in Oregon. Major General Bentz holds a PhD and master of science in nuclear engineering from the University of Missouri, a master of science in national security strategy from the National Defense University, and a bachelor of arts in radiological health from Oregon State University. She published the article "Learning from Catastrophe: Lessons from the COVID-10 Pandemic for Preparing for and Responding to a Domestic Radiation Emergency" (2021, an NTI paper) that addresses insights and lessons related to preparedness for other major disasters.

Michael Dunning retired from Lawrence Livermore National Laboratory (LLNL) in 2018, where he was principal deputy principal associate director of Weapons and Complex Integration. Prior to this role, Dunning was the program director and division leader for Primary Nuclear Design (PND). Under his leadership, PND operated a large-scale experimental facility at Site 300 and LLNL's High Explosives Applications Facility—both of which are national resources for the study of high explosives, conventional munitions, and propellants. PND also maintained a vigorous code-development and simulation capability, using some of the world's most capable supercomputers. During his career leading up to the leadership roles in the weapons community, Dunning was recognized as a weapons designer responsible for the first subcritical experiments for LLNL (1994–1999), was the LLNL representative to the Department of Energy (DOE) Defense Programs "Science Council" in Washington, DC (1998), and worked with the National Nuclear Security Administration in the DOE help establish a counter nuclear terrorism program focused on understanding novel potential threat devices (1999–2000). Prior to 2006, he also served as the Nevada Experiments and Operations program leader and was responsible for the LLNL program and operations executed at the Nevada Test Site (NTS). Dunning also led LLNL efforts to detect, assess, and disable unauthorized nuclear and radiological dispersal devices. Dunning joined the laboratory in 1989 as a postdoctoral researcher. He completed his bachelor's, master's, and doctoral degrees in nuclear engineering at the University of Michigan. Since retiring, Dunning has participated in several reviews supporting the Associate Administrator for Counter-terrorism & Counter-proliferation (NA-80) and was part of a small team compensated for reviewing the Joint Technical Operations Team in the summer of 2023.

Robert C. Dynes is a professor of physics at the University of California, San Diego (UCSD). He was the 18th president of the University of California, from 2003 to 2008. A first-generation college graduate and a distinguished physicist, Dr. Dynes served as the sixth chancellor of UCSD's campus from 1996 to 2003. His numerous scientific honors include the 1990 Fritz London Award in Low Temperature Physics and his election to the National Academy of Sciences in 1989 and the American Academy of Arts and Sciences in 1994. As a professor of physics at UCSD, he founded an interdisciplinary laboratory where chemists, electrical engineers, and private industry researchers investigated the properties of metals, semiconductors, and superconductors. He subsequently became chair of the Department of Physics and then senior vice chancellor for academic affairs. Since leaving the UC presidency in June 2008, Dr. Dynes has joined the boards of Argonne National Laboratory, the review panel for the Canadian Foundation for Innovation, the Helmholtz Foundation in Germany, and the San Diego Foundation. For the National Academies of Sciences, Dr. Dynes is currently the co-chair for the Intelligence Community Studies Board, chaired three consensus studies (Disposal of Surplus Plutonium in the Waste Isolation Pilot Plant, Improving the Assessment of Proliferation Risk of Nuclear Fuel Cycles, and Evaluating Testing, Costs, and Benefits of Advanced Spectroscopic Portals), and was a committee member for Protecting Critical Technologies for National Security in an Era of Openness and Competition and Assessment of Inertial Confinement Fusion (ICF) Targets. The study's report on advanced radiation detectors was given to the Department of Homeland Security in March 2009. A native of London, Ontario, Canada, and a naturalized U.S. citizen, Dr. Dynes holds a bachelor's degree in mathematics and physics and an honorary doctor of laws degree from the University of Western Ontario, and master's and doctorate degrees in physics and an honorary doctor of science degree from McMaster University. He also holds an honorary doctorate from l'Université de Montréal. Currently, Dr. Dynes is consulting for Decision Sciences, a company developing muon sensors capable of detecting high Z materials including special nuclear materials.

Steven A. Fetter is an associate provost, dean of the graduate school, and professor of public policy at the University of Maryland. He is a member of the Council on Foreign Relations and a fellow of the American Physical Society. Dr. Fetter worked for five years in the White House Office of Science and Technology Policy during the Obama Administration, leading both the National Security and International Affairs and the Environment and Energy divisions. In 1993-1994, he served as special assistant to the assistant secretary of defense for International Security Policy and he worked in the State Department as an American Institute of Physics fellow. He has been a member of the director of national intelligence's Intelligence Science Board and DOE's Nuclear Energy Advisory Committee; served as president of the Association of Professional Schools of International Affairs and as vice chairman of the Federation of American Scientists; and has been a visiting fellow at Stanford's Center for International Security and Cooperation, Harvard's Center for Science and International Affairs, MIT's Plasma Fusion Center, and Lawrence Livermore National Laboratory. Dr. Fetter received a bachelor of science in physics from the Massachusetts Institute of Technology in 1981 and a PhD in energy and resources from the University of California, Berkeley, in 1985.

William Goldstein retired in 2021 after serving 7 years as the 12th Director of Lawrence Livermore National Laboratory (LLNL). In this role, he had responsibility for all aspects of research, administration, and operations for ~7500 staff and a budget of ~$2.5B. As LLNL Director, Goldstein submitted a letter annually to the President assessing the safety, security and reliability of the US nuclear stockpile. Dr. Goldstein is a fellow of the APS and AAAS, and the recipient of four Certificates of Excellence awarded by the Secretary of Energy, including one for his support to the successful negotiation of the JCPOA. As LLNL Director, Goldstein led LLNL as the hosting site for APEX Gold, an international symposium on nuclear incident response, and later participated personally in APEX Blue, a smaller exercise involving Israel and the US. Goldstein has published over 70 peer reviewed papers in many fields, but primarily in plasma spectroscopy and the physics of highly charged ions. His most significant work was presented in a series of papers on atomic transition arrays in plasmas, which led to the Super Transition Array scheme for simulating radiative opacities in hot, dense plasma. Goldstein holds a PhD in theoretical physics from Columbia University and a BA in physics from Swarthmore College.

Edward B. Held served for three decades as a clandestine operations officer of the Central Intelligence Agency. Following his retirement from CIA, Mr. Held served as Chief of Counterintelligence at Sandia National Laboratories where he earned the trust and support of the Sandia scientific community to implement a cost effective program to aggressively counter the activities of foreign intelligence services targeting the laboratory. Impressed by this successful strategy, Energy Secretary Steven Chu asked Mr. Held to return to federal service as Director of DOE's Office of Intelligence and Counterintelligence to lead all intelligence and counterintelligence activities at DOE and the DOE-owned national laboratories. Secretary Chu was particularly appreciative of Mr. Held's success in expanding DOE strategic influence on national cyber defense policy. Director of National Intelligence James Clapper awarded Mr. Held the National Intelligence Superior Service Medal for "transforming the Office of Intelligence and Counterintelligence to meet threats to the national security of the United States." In June 2013, Energy Secretary Ernest Moniz asked Mr. Held to return to federal service again and assume the responsibilities of DOE Associate Deputy Secretary as well as Acting Undersecretary for Nuclear Security with responsibility for leading the National Nuclear Security Administration, a $12 Billion dollar enterprise with 35,000 government and laboratory employees. Secretary Moniz praised Mr. Held's leadership of NNSA as "transformational." He is currently retired.

Eleanor Melamed retired in June 2020 from the position of associate assistant deputy administrator for the Office of Global Material Security (GMS) in the National Nuclear Security Administration. The purpose of this office was to prevent terrorists from obtaining and using nuclear and radiological materials. This work was carried out by working with partners worldwide to build sustainable capacity to secure these materials, and to interdict and investigate the trafficking of those materials. Melamed joined the Department of Energy in 1994, where she worked for the DOE Office Environment, Safety and Health on projects in the United States and Russia related to radiation exposure. She moved to the National Nuclear Security Administration Office of Defense Nuclear Nonproliferation in 2000, where she has worked on a variety of nuclear nonproliferation issues. Beginning in July 2003, Melamed worked in the Office of the Second Line of Defense, recently renamed Nuclear Smuggling Detection and Deterrence (NSDD). In 2005, she was appointed deputy director of the Program and in October of 2011, she became the director. She was instrumental in promoting the growth of NSDD from a start-up program with a small federal staff, and one foreign partner, to a program of recognized effectiveness and global reach. Before joining this program, Melamed was deputy director and then acting director of the Nuclear Cities Initiative, a joint U.S./Russian program designed to assist the Russian Government in developing sustainable non-weapons employment for its weapons scientists. She is currently a board member for the Nuclear and Radiation Studies Board at the National Academies of Sciences. Melamed holds a bachelor of arts from Oberlin College and a master of arts in European History from the University of Chicago.

Brendan G. Melley is the director of the Center for the Study of Weapons of Mass Destruction (CSWMD), Institute for National Strategic Studies, at National Defense University. He joined CSWMD as a senior research fellow in 2011, and his civilian government experience includes senior staff assignments at the Defense Intelligence Agency, the President's Foreign Intelligence Advisory Board, and on the National

Security Council staff as director for both Proliferation Strategy and for Intelligence Programs. At the NSC, he supported priority U.S. intelligence and counterproliferation efforts and led the development and coordination of weapons of mass destruction (WMD) interdiction policies across the federal government and with foreign partners. As director, Melley oversees CSWMD's three lines of work: research and analysis involving the role of WMD in U.S. national security policies and strategies, including deterrence and countering of WMD plans and activities; direct policy support to senior defense and other USG leaders on WMD-related matters; and WMD education within the Joint Professional Military Education system and to other academic institutions in the Department of Defense. Melley served on active duty in the U.S. Army as a light infantry and military intelligence officer. He graduated from Providence College, Rhode Island, the Postgraduate Intelligence Program at the Defense Intelligence Agency, and earned a master of science in WMD Studies from Missouri State University.

Scott Roecker is the vice president for Nuclear Material Security at the Nuclear Threat Initiative (NTI). He focuses on work to reduce the risk associated with nuclear and radiological materials through a number of NTI projects, including the Global Dialogue, Nuclear Security Index, and targeted cooperation with priority countries. He also supports nuclear verification efforts and advises leadership on issues related to Iran. Roecker previously served as the director of the Office of Nuclear Material Removal at the National Nuclear Security Administration (NNSA). In this position, he was responsible for all U.S.-led activities to remove or dispose of highly enriched uranium (HEU) and separated plutonium with partner countries. From 2014 to 2017, Roecker served as the director for Nuclear Threat Reduction at the National Security Council. In that role, he coordinated the development and implementation of policies to prevent nuclear terrorism through efforts to secure nuclear and radiological materials, minimize HEU and plutonium, and counter nuclear smuggling. He also was the Sous-Sherpa for the United States at the 2016 Nuclear Security Summit. Roecker holds a masters in international affairs from The George Washington University and a bachelor of science from the University of Minnesota. Mr. Roecker published the article "Nuclear Power Plants under Attack: The Legacy of Zaporizhzhia" (April 2023 in *Arms Control Today*) that discusses the international response to events at the Zaporizhzhia Nuclear Power Plant.

Jessica Stern is a research professor at Boston University and a senior fellow at both the Center for Naval Analyses and the Community Safety Branch at the Harvard T. H. Chan School of Public Health. Stern has taught courses on counterterrorism for over 20 years. She is a member of the Homeland Security Experts Group. Stern is the coauthor with J. M. Berger of *ISIS: The State of Terror*; and the author of *My War Criminal, Denial, Terror in the Name of God* and *The Ultimate Terrorists*. Stern served on President Clinton's National Security Council Staff in 1994-1995. She was included among seven "thinkers" in *Time Magazine*'s 2001 series profiling 100 innovators. She was selected as a Guggenheim Fellow in 2009, a World Economic Forum Fellow from 2002 to 2004, and an International Affairs Fellow in 1994, and she was elected to Sigma Xi, an engineering honors society, in 1986. Stern advises a number of government agencies on issues related to terrorism. She has a bachelor's degree from Barnard College in chemistry, a master's degree from MIT in technology policy (chemical engineering), and a doctorate from Harvard University in public policy. She is a 2016 graduate of the Massachusetts Institute of Psychoanalysis.

Kristine L. Svinicki currently serves as adjunct professor of nuclear engineering and radiological sciences in the College of Engineering at the University of Michigan. She is an internationally recognized policy expert and innovator with over 30 years of public service at the state and federal levels. Appointed to her role by three successive United States presidents, Svinicki is the longest-serving member of the United States Nuclear Regulatory Commission (U.S. NRC) in the agency's history, having stepped down as chairman in early 2021. Prior to her appointment to the U.S. NRC, Svinicki served as an expert and policy advisor for over a decade to members of the United States Senate on subjects ranging from energy to national security. She previously managed nuclear research and development programs at the U.S. Department of Energy and worked as an energy analyst for the State of Wisconsin. She was selected as a Brookings Institution Fellow in 1997 and as a John C. Stennis Congressional Fellow of the 108th U.S. Congress. She currently sits on the boards of TerraPower, the Southern Company, and Pinnacle West Capital Corporation and serves on the Idaho National Laboratory Nuclear Science and Technology Strategic Advisory Committee. Svinicki was awarded a bachelor of science in Nuclear Engineering from the University of Michigan in 1988.

Rodney K. Wilson retired from Sandia National Laboratories in March 2020, where he was the director of Sandia's Center for Global Security and Cooperation that supports U.S. government agencies responsible for nonproliferation and arms control, cooperative threat reduction, and international security. His other assignments included director for National Security Studies & Integration, and chief of staff to the Deputy Laboratories director for Nuclear Weapons Programs. Dr. Wilson began his career at Sandia in 1980, and during nearly 40 years at Sandia led and participated in a broad range of national security–related activities, including: the security and survivability of theater nuclear forces; studies of nuclear weapon transportation safety and security; studies on the transparency and verification of nuclear warhead dismantlement; and activities that support government programs to prevent the misuse of nuclear, chemical, biological, and radiological materials. He also worked in the Office of Policy Planning, Analysis and Assessment at the National Nuclear Security Administration (NNSA) when it was created in 2001. Upon his retirement in 2020, he was awarded the NNSA Administrator's Distinguished Service Gold Award. Dr. Wilson holds a bachelor of science in engineering (BSE) in engineering science from the University of Michigan, and both a master of science and PhD in theoretical and applied mechanics from the University of Illinois.

TECHNICAL CONSULTANT

Nickolas Roth serves as a senior director on the Nuclear Threat Initiative's Nuclear Materials Security Program team, where he focuses on reducing the risk of nuclear terrorism, advancing arms control and disarmament, and strengthening institutions that support nonproliferation. Roth earned a master of public policy from the University of Maryland. He is currently a research scholar at the Center for International Security Studies at the University of Maryland and an associate of the Project on Managing the Atom at the Harvard Kennedy School.

STUDY DIRECTOR

Michael Janicke joined the National Academies as a senior program officer on the Nuclear and Radiation Studies Board from Los Alamos National Laboratory. Dr. Janicke graduated from Rice University with a B.S. in chemical engineering and continued his education at the University of California, Santa Barbara, where he earned his PhD in chemical engineering. Following his studies, Dr. Janicke was an Alexander von Humboldt Fellow at the Max Planck Institute for Carbon Research in Mülheim an der Ruhr. While in Germany, he worked with Professor Ferdi Schüth, former vice president of the German National Science Foundation. In 2000, Dr. Janicke returned to New Mexico as a postdoctoral fellow at Los Alamos National Laboratory (LANL) and became a staff member in 2002. Most recently he was the center director for REFOCUS, the Resonance Center for Chemical Signatures, and spearheaded efforts in developing new methods to detect chemical threat agents and synthetic opioids at border and airport checkpoints using magnetic resonance techniques. At LANL he was also involved in several programmatic studies for Enhanced Surveillance Campaigns and Lifetime Extension Programs for the weapons community, participated in NA-22 projects analyzing funded research programs across the U.S. Department of Energy complex, and assisted in addressing chemical questions associated with Medical Isotope and Basic Energy Sciences Heavy Element programs.